MANAGING URBAN WATER SUPPLY

Water Science and Technology Library

VOLUME 46

The titles published in this series are listed at the end of this volume.

MANAGING URBAN WATER SUPPLY

Edited by

DONALD E. AGTHE

Consultant and retired Professor,
Tucson, Arizona, U.S.A.

R. BRUCE BILLINGS

Department of Economics,
University of Arizona, Tucson, U.S.A.

and

NATHAN BURAS

Department of Hydrology and Water Resources,
University of Arizona, Tucson, U.S.A.

Springer-Science+Business Media, B.V.

A C.I.P. Catalogue record for this book is available from the Library of Congress.

ISBN 978-90-481-6470-7 ISBN 978-94-017-0237-9 (eBook)
DOI 10.1007/978-94-017-0237-9

Printed on acid-free paper

Table of Contents

Table of Tables

Table of Figures

Preface

We, the editors, have long believed that a strong knowledge of relatively simple economic and engineering concepts is valuable in solving water management problems. The lack of such knowledge has been apparent to us in some of the journal articles, research proposals and books we have reviewed. The articles which have been written concerning specific local water economics and management issues are scattered over a wide variety of journals, making them hard to access. Most of the extensive water resources literature is concerned with large regional water projects or with narrow technical and regional issues.

This book was written to make practical economic and engineering concepts readily available to urban water supply managers, thereby filling a gap in the available literature. It is concerned with decisions made daily, monthly, or annually by managers of urban water supply systems. The book includes basic chapters presenting supply and cost concepts, calculation of demand elasticities, use of marketing concepts, public goods analysis, water markets, industrial water demand and the use of price in water conservation. The authors have included multiple examples of how these concepts can aid in managing urban water supply.

The water provider is generally a governmental entity or regulated private utility. Most books on public utilities and their management emphasize gas, electricity, or telephone rather than water. Water is different because of major variations in quality by source and the necessity for proper disposal of waste water.

We focus exclusively on the problems faced by the water utility and include examples of how elementary economic principles can be applied to make the water utility more *efficient* and more *equitable* in its operation. In some situations, increased efficiency can be achieved only by a sacrifice of equity, thereby requiring painstaking analysis. While various professionals can generally agree regarding actions that would improve efficiency, they often disagree over issues of equity. This book devotes considerable effort to show how these two concepts can be applied in developing management, pricing, and conservation policies for an urban water provider.

The editors designed this book as a comprehensive reference to economics and management applied to urban water provision. It is devoted to the policy application aspects of economics and management with only limited "as necessary" discussion of programming and econometric measurement techniques. Those interested in detailed econometric, programming, or forecasting techniques should look elsewhere as there are several very excellent texts available in these subjects.

The organization of the book is determined by our purpose to present the material so that it will be accessible to water resource managers who may have limited training in economics. Chapters two through seven introduce the principles of economics, hydrology, and engineering to the reader. Topics such as rate structures, demand and supply determinants, derived demand, economics of new supply sources, economic efficiency and equity are discussed in these chapters along with relevant policy applications. The later chapters, written by recognized specialists in their fields, include individual case studies of Israel and the United Kingdom as well as detailed analysis of water supply, water markets, subsidence, industrial demand and drinking water protection.

We especially wish to thank each of the authors who contributed their time and expertise to make this book possible. We anticipate that our readers will benefit from their knowledge as we have. We also would like to thank George Young, meteorologist at Pennsylvania State University; Staci Warne, our word processing guru; Judith Billings, who proof read the entire book; Petra van Steenbergen and Martine van Bezooijen, our editors for their valuable comments. Of course, no work is without errors and they remain our responsibility.

Donald E. Agthe
R. Bruce Billings
Nathan Buras
Tucson, Arizona, 2003

Contributors

Donald E. Agthe

Economic Consultant
2509 N. Campbell Ave., PMB 133
Tucson, Arizona 85719 U.S.A.

Dr. Agthe has a Ph.D. from Florida State University in Economics and an M.S. in Agricultural Economics from Rutgers University. He is the author of over 50 publications including refereed articles in *Land Economics, Water Resources Research, Water Resources Management, Journal of the American Water Resources Association, Journal of Water Resources Planning and Management, Policy Studies Review, The American Economist, and the American Journal of Economics and Sociology.* While currently a consultant, his over 25-year career in academia included faculty appointments in economics at Arizona State University, St. Mary's University of San Antonio, University of Nevada, and Central Michigan University.

R. Bruce Billings

Associate Professor of Economics
Department of Economics
McClelland Hall 401
University of Arizona
Tucson, Arizona 85721U.S.A.

Dr. Billings has a Ph.D. from Claremont University Graduate School and an M.A. from The University of Arizona in Economics. He is the author of over 50 publications including refereed articles in *Land Economics, Journal of Regional Science, National Tax Journal, Water Resources Research, Water Resources Management, Journal of the American Water Resources Association, Journal of Water Resources Planning and Management, The American Economist, and the American Journal of Economics and Sociology.* Professor Billings is co-author of *Forecasting Urban Water Demand* and *Current Regional Issues--Arizona, Colorado, Nevada, New Mexico, Oklahoma, Utah.* He has held consulting positions with the U.S. Bureau of Reclamation, the State of Hawaii, the State of Arizona, and the Idaho National Engineering Laboratory. He has served as an outside reviewer of grant applications for the Universities Council on Water Resources for the United States Geological Survey and for the Water Resources Institute. His biography appears in Marquis Who's Who in America.

Nathan Buras

Professor of Hydrology
Department of Water Resources and Hydrology
University of Arizona
Tucson, Arizona 85721 U.S.A.

In addition to a long career at the University of Arizona that includes nine years as department chairman, Dr. Buras has served on the faculties of Stanford University and the University of Tokyo. He holds a Ph.D. from UCLA and a M.S. from Techunion, Haifa, Israel. Professor Buras is author or co-author of over 100 scientific and professional publications as well as over 40 reports and other publications related to his consulting and professional work. A contributor of chapters in more than 10 books, he is the editor of *Management of Water Resources in North America III, Anticipating the 21st Century* and *Reflections on Hydrology – Science and Practice* as well as the author of *Scientific Allocation of Water Resources.* He is a Fellow of the American Society of Civil Engineers and the International Water Resources Association.

Kenneth F. Clarke

Water Engineering and Management Consultant
Fallowfields, Glatton, Huntingdon
Cambridgeshire, PE28 5RU, United Kingdom

Holder of a M.S. in water resources technology from Birmingham University [U.K.], he is a fellow of the Institution of Civil Engineers [U.K.], a fellow of the Chartered Institute of Management [U.K.], and a fellow of the Royal Geographical Society [U.K.]. Mr. Clarke has held professional positions with Oxford City Council, Trent River Authority, and the Water Resources Board for England and Wales. He was assistant director of the Anglian Water Authority and the Director of Operations of the Vivendi owned Three Valleys Water Company. While he was with the Water Resources Board, engaged in national water resources planning, he developed a mathematical model to carry out appraisals of national water resource development options. In 1993-1994 he was president of the Chartered Institution of Water and Environmental Management, an organization of over 12,000 members in 40 countries. Since 1991, Mr. Clarke has been a private consultant to the United Kingdom, Egypt, Uzbekistan, Australia and India.

Bonnie G. Colby

Professor of Agricultural and Resource Economics
Department of Agricultural and Resource Economics
The University of Arizona
Tucson, Arizona 85721U.S.A.

Dr. Colby holds a Ph.D. from the University of Wisconsin. The leading journals for her publications include *The American Journal of Agricultural Economics, Water Resources Research, Land Economics,* and the *Journal of Environmental Economics and Management.* She has contributed chapters to four books. Her expertise is in the economics of natural resource policy and disputes over water, the public lands and environmental regulation. She has provided invited testimony on these matters to state legislatures and to the U.S. Congress. She served on the National Research Council's Committee on Western Water Management and currently is serving on the National Academy of Science committee investigating use of economic methodology by the Army Corps of Engineers, involving billion-dollar proposed projects on U.S. waterways. Dr. Colby advises public and private sector organizations on managing natural resources and resolving environmental disputes.

Laura Coplin

U.S. Geological Survey
Austin, Texas 78757U.S.A.

Ms. Coplin holds a B.A in Geology from Rice University. She began work for the United States Geological Survey in 1978 serving the Texas District of the Water Resources Division as hydrologist, hydrologic technician, and scientific illustrator in the publications unit. She worked in the Houston sub-district from 1978 to 2002, and served most recently as project chief, ground water data chief, and co-site administrator. She is a member of the Subsidence Interest Group of the U.S.G.S. She has authored numerous U.S.G.S. publications and co-authored the first technical publication of the Harris-Galveston Coastal Subsidence District.

Benedykt Dziegielewski

Professor of Water Resources
Department of Geography
Faner Hall 4531, MS 4514
Southern Illinois University
Carbondale, Illinois 62901 U.S.A.

Dr. Dziegielewski holds a Ph.D. in Geography and Environmental Engineering (Interdisciplinary Degree), from Southern Illinois University, and an M.S. in Environmental Engineering from Wroclaw Polytechnic University in Poland. In 1984 the Universities Council on Water Resources Inc. presented him with an award for the Outstanding Water Resources Thesis. He is the Executive Director of the International Water Resources Association. His research interests include water demand management, modeling of water use, and drought management. He has significant journal publications in *Water Resources Bulletin* and *Environment* and is co-author of several chapters in books. Dr. Dziegielewski is co-author of *Committee on U.S.G.S. Water Resources Research. 2001: Development of the National Water Use Information Program* and co-author of three American Water Works Association publications: *Residential End Uses of Water, Commercial and Institutional End Uses of Water* and *Evaluation of Urban Water Conservation Programs: A Procedures Manual.*

Devin L. Galloway

Western Regional Ground-Water Specialist
U.S. Geological Survey
Sacramento, California 95826 U.S.A.

Devin Galloway received an M.S. degree in Environmental Science from Indiana University and an MS degree in Civil and Environmental Engineering from the University of Illinois. He has been a member of the U.S.G.S. since 1978 and is currently the Western Regional Ground-Water Specialist, Water Resources Division. He is the principal editor, with Steve Ingebritsen and David Jones, of *Land Subsidence in the United States* (U.S.G.S. Circular 1182, 1999). He has authored or coauthored many articles on land subsidence, including several recent (1998-02) papers in the peer-reviewed literature on the application of space-based InSAR (interferometric synthetic aperture radar) to basin-scale studies of aquifer-system compaction related to ground-water pumping. Galloway is a member of the UNESCO IHP Working Group on Land Subsidence.

Steve Ingebritsen

Chief of the Branch of Regional Research
U.S. Geological Survey
345 Middlefield Road
Menlo Park, California 94025U.S.A.

Dr. Ingebritsen holds a Ph.D. in Hydrogeology from Stanford University. He has been a member of the U.S.G.S. since 1980. He is author, with Ward Sanford, of *Groundwater in Geologic Processes* (Cambridge University Press, 1998). Ingebritsen received the "John Hem Excellence in Science and Engineering Award" from the National Groundwater Association in 2000 and served as the "Birdsall-Dreiss Distinguished Lecturer" for the Geological Society of America in 2001.

Yoav Kislev

Professor
Department of Agricultural Economics and Management
Hebrew University of Jerusalem
Rehovot,76100 Israel

Professor Kislev holds a Ph.D. in Economics from the University of Chicago and a M.S. in Agriculture from Hebrew University. He has refereed publications in *The American Journal of Agricultural Economics, The Australian Journal of Agricultural Economics, Journal of Political Economy, Economic Development and Cultural Change, Journal of Development Studies, Metroeconomica,* and the *Journal of Economic History.* In addition to contributing chapters to over 10 books, Dr. Kislev has been editor, co-editor, and co-author of books including *Agricultural Research and Productivity; An Economic Analysis of Flood Control Projects in the Hula Valley; A Supply Cooperative of Moshavim: Plants, Economic Functioning, and Finance; Agricultural Cooperatives in Transition;* And *Technology Cooperation, Growth and Policy—Studies in Agricultural Economics.* He has also served as a consultant to the World Bank.

Mark A.R. Kleiman

Professor of Policy Studies
School of Public Policy and Social Research
University of California, Los Angeles
Los Angeles, California 90095U.S.A.

Dr. Kleiman, holder of a Ph.D in Public Policy from the John F. Kennedy School of Government at Harvard University, is Professor of Policy Studies at the University of California at Los Angeles. His specialties include drug policy, crime control, and application of models of imperfect rationality to public policy. He is author of books including: *Against Excess: Drug Policy for Results* and *Marijuana: Costs of Abuse, Costs of Control* and numerous journal articles and chapters in books.

Ronnie Levin

Visiting Research Scientist
Harvard University School of Public Health
Water and Health Program
Boston, Massachusetts 02115U.S.A.

Ronnie Levin has been s a Senior Scientist at the U.S. Environmental Protection Agency and is currently working in the Water and Health Program at the Harvard School of Public Health. Previous positions at EPA have included Staff Director for the Water Staff of the Office of Research and Development in Washington, D.C. and Economic Analyst in the Office of Policy Analysis in Washington D.C. She holds an M.A. from Brandeis University. She has published chapters in a least five books and articles in the *American Journal of Public Health, Epidemiology, Environmental Health Perspectives, International Journal of Epidemiology and Public Health, Environmental Research* and *International Journal of Epidemiol and Community Health.* Her research specialties include quantitative risk assessment, cost-benefit analysis and human health effects of drinking water quality.

Steven Renzetti
Professor of Economics
Department of Economics
Brock University
St. Catherines
Ontario, Canada L2S 3A1

Professor Renzetti holds a Ph.D. in Economics from the University of British Columbia. He has contributed chapters to several books and is author of *The Economics of Water Demands* and editor of *Economics of Industrial Water Use*. His work has appeared in refereed journals such as *Environmental and Resource Economics, Canadian Journal of Economics, Canadian Journal of Agricultural Economics, Land Economics, Water Resources Research,* and *Journal of Environmental Economics and Management*. Many of his presentations at conferences in North America and Europe have appeared in refereed conference proceedings. He has received refereed external grants from the Donner Foundation, Environment Canada, and Eco-Research Tri-Council among others. Consulting clients have included the Ontario Ministry of Finance, Ontario Ministry of Natural Resources, Environment Canada, and The World Bank. He was appointed to the Advisory Panel of the Canada-Ontario Use and Supply Project and to the International Joint Commission's Lake Ontario St. Lawrence River Water Levels Study Board.

Chapter 1

Introduction

Donald E. Agthe
Economic Consultant
R. Bruce Billings
University of Arizona
Nathan Buras
University of Arizona

1.1 Introduction

Urban water supply is a growing issue for many cities and towns. Rapid population growth, drought, ground water contamination, surface water pollution, greater use of water using devices as a result of higher incomes and ground water mining have all led to water shortages in urbanized areas. This book examines how urban water supplies can be managed efficiently while satisfying community standards of equity. It includes economic, managerial, and hydrologic analysis of water pricing, selection of water rate structures, wastewater use, when and how to add new sources of water into the current system, and non-price conservation measures.

A simple model relating water demand, cost and availability is shown in Figure 1.1. It is clear that cost influences both the availability and demand for water. Demand affects the long run availability of water because as demand increases people seek new supplies of water by building dams, developing new well fields, and/or contracting with more distant sources. Availability affects water demand because people are not likely to locate water-using industrial activities in areas of limited water supplies.

D.E. Agthe et al. (eds.), Managing Urban Water Supply, 1–9.

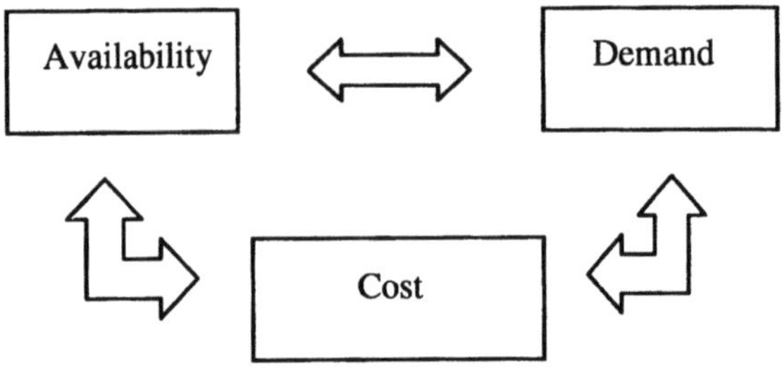

Figure 1.1 Water Use Cycle

1.2 Economists and Engineers

Many urban water supply decisions are made directly by, or with the advice of, engineers and economists. In his book, *Economic Analysis at EPA*, Richard D. Morgenstern (1997) says that these actors do not think alike when analyzing problems or developing policy. He points out that engineers are trained to solve problems rather than to resolve them. They expect to find the correct answers by working with rigid formulas. Economists, on the other hand, tend to emphasize trade-offs and prefer to examine a range of possible choices. Furthermore, technological efficiency and economic efficiency may lead to different quantities of water being provided since these concepts differ in their underlying assumptions. As Nobel Laureate George Stigler (1966) points out, "the choice of production methods must take account of the importance of the inputs as well as that of the finished goods." Richard Leftwich (1966) uses the example of an internal combustion engine being more efficient mechanically than a steam engine but if the steam engine uses sufficiently cheaper fuel in a particular application, it could be more economically efficient.

In urban water supply management, engineers are likely to seek technological solutions such as new dams, additional well fields, or elaborate water importation schemes to increase the water supply to meet increasing quantities of water desired by the public. They see their mission as providing the quantity of water the public wants. The economist examines the incremental costs of obtaining the additional water, the incremental benefits of using this water, and may suggest raising the price of water to eliminate some or all of the extra wants. The economist, in the interest of equity or profit maximization, might suggest separate markets for different classes of water, special rate structures to focus on a particular demand segment or special programs to provide water for low income families. Thus, the economist and the engineer may have quite different ideas about the optimum quantity of water to be delivered. The engineer views quantity demanded as the sum of all the water that people want while the economist views the quantity demanded as being subject to the price of water and the quantity supplied as being subject to its production, treatment and delivery cost.

While economists label economics as a science because it follows the scientific method, economists face a problem that engineers can usually avoid. Their subjects' actions are affected by the conduct of their research. Stigler (1966) provides an excellent illustration of this problem:

> Imagine the problem of a chemist if he had to deal with molecules of oxygen, each of which was somewhat interested in whether it was joined in chemical bond to hydrogen. Some would hurry him along; others would cry shrilly for a federal program to drill wells for water instead; and some would blandly assure him that they were molecules of argon. And this chemist, who in analogy would also be a chemical element, could never be absolutely certain that he was treating other elements fairly. Several elements would hire their own chemists to protect their interests

Economists view shortages as a market phenomenon resulting from a failure to raise prices. If the quantity available is less than the quantity demanded, price could be increased sufficiently to clear the market. However, higher water prices may be politically unacceptable or be viewed as inequitable. Thus, communities may choose to augment price rationing with non-price measures such as sprinkling restrictions or requirements for low water use appliances.

This book is edited by two economists, Agthe and Billings, and an engineer, Buras. Most likely, it reflects the biases of the two groups represented. However, we show that economists and engineers can work together to develop good solutions to urban water problems; an idea expressed for other water related policies by Boulding (1964). Combined economic and engineering solutions to urban water supply problems are superior to those of either discipline alone. This work emphasizes demand management as equal in importance to supply considerations in developing urban water policies.

1.3 Economic Quantity of Water

The economically useful quantity of water available depends upon its quality. The total economic quantity of water increases as the quality of water required by the proposed end use gets lower because the quantity available at any use level includes the water available of that quality level plus the total water of all quality levels above it.

The highest quality water is that needed for drinking and certain high-tech or very clean production processes. Lower quality water is suitable for most industrial and recreational uses so it may be economically inefficient to use drinking quality water for these purposes. For example, treated effluent or minimally treated river water can be used for irrigating landscapes and golf courses. Water should be priced according to its quality to encourage use of lower quality water when it is available. Potable water should be priced much higher than reclaimed waste water, for example. A lower price would encourage the full utilization of reclaimed wastewater and reflect its lower economic value. A subsidy may be required to offset the cost of the extensive treatment of wastewater required before it can be reused and the expense of developing a separate delivery system. A subsidy would also aid in developing a sufficiently large market to achieve economies of scale in treatment and delivery of reclaimed water.'

1.4 Demand, Prices, Rate Structures

Demand Curves show the relationship of water used to its price. While there is no substitute for water to meet daily minimum household requirements for drinking, cooking, and sanitation, these account for only about 11 percent of total urban water use in the U.S. (Johnson, 1988). Most household water use is derived from an underlying demand for clean clothes and cars, gardens, decorative fountains and swimming pools. Other derived demands result from a desire to play (parks, golf courses, football fields) or from production processes such as cleaning and cooling. Market demand for water is the sum of these derived demands and the direct demands for household water.

The most frequently used measure of the response of water use to price is the *price elasticity of demand*, the percentage change in the quantity of water purchased for a one percent change in its price. Howe and Linaweaver (1967) and Gottlieb (1963) were among the first to empirically estimate the price elasticity of demand for water. Their results clearly showed that water use fell in response to higher prices. In contrast, Wong (1972) found no significant quantity response to water price in Chicago. He concluded that the price of water was too low for price to be an effective rationing device. However, as water became scarce and its price increased, researchers found an inverse relationship between price and water use. For example, Martin and Thomas (1986) compared water use among several high income desert-climate cities with very different prices, finding an overall price elasticity of demand of -0.5. These results are similar to those of Agthe and Billings (1980) and Foster and Beattie (1979). These results indicate that the demand for urban water is price inelastic, so that total revenue to a local water utility will increase when the price is raised. Therefore, from a strictly economic or engineering viewpoint, additional revenue can be readily obtained to finance increased costs associated with conservation efforts or supply expansion.

In general, water has substitutes. Capital and labor substitutes in response to a water price rise may include leak inspection and replacement programs, more efficient water using appliances, shorter showers, lower water use landscapes and irrigation methods, recycling, and better swimming pool management. Substitutes such as these exist within all urban markets for water - commercial, industrial, government, apartments, individual residences, and recreational. The voluntary adoption of non-price technical or behavioral conservation measures are frequently a response to price increases (Agthe and Billings, 1997). Technical responses include the installation of water saving devices such as low flow shower heads, faucets, and small flush toilets, while behavioral changes include shorter showers, turning off faucets between uses, and less frequent irrigation of landscapes

It was suggested above that water price can be an effective tool in demand management. Setting prices to limit demand rather than to simply cover cost is often referred to as *conservation pricing*. While this could be accomplished by simply setting a single high price per gallon for water, concerns for equity typically lead to a more complex rate structure. Typical single family residential conservation rate structures include a low price for the first few units of water, a somewhat higher price for the next several units, and a much higher price for all additional units. Such a multiple increasing block rate structure provides incentives for conservation by high volume users while providing low water bills to those using relatively small amounts. The concept behind increasing block rate conservation pricing is that the largest volume users have more opportunities to save water and that their marginal uses have less value to society than those of the smaller volume users. Also underlying this rate structure is the concept that larger volume users tend to have higher incomes making water a smaller portion of their budget than for the lower income smaller volume users. Given an identical increase in water price, the largest volume users would be less likely to decrease their water use in response to the price change than smaller volume users and, therefore, the large volume water users must face a higher price as an incentive to reduce their use appropriately. Numerous studies have shown that water use increases with income, so these conservation rates are not a burden on those with already low incomes.

Simple economic efficiency in water use requires that all water be priced at its marginal cost. This implies a uniform price for all customers, and abandons the potential gains in equity that may result from the block rate pricing structure described above. A variant of this "one flat rate for all customers" is a two-part tariff in which the price per unit of water is equal to its marginal cost and a periodic availability of service fee is assessed to cover fixed costs associated with water delivery. However, setting price equal to marginal cost does not guarantee that the utility will break even. It may collect either more or less revenue than required to cover all costs. Depending on the individual utility, conservation rates (of the flat rate or increasing block varieties) sufficient to reach some community goal may generate substantial surpluses. This is often politically unacceptable, preventing the enactment of such rates.

An urban water market represents a collection of sub-markets: household, commercial, industrial, governmental, and recreational. Each sub-market has its own demand characteristics with reference to quality requirements, peak load volume, production functions for secondary goods and services, and ability to substitute other inputs such as capital and labor for water. The waste water stream will also vary by sub-market.

1.5 Long Run and Short Run Water Supply

Water supplies are typically analyzed on either a short run basis, concerning the best utilization of existing capital, or a long run basis, focused on the potential need for expanded capacity. In the short run, the marginal cost (the extra cost of one more unit) of adding one more customer to an existing distribution system or providing a incrementally larger volume of water is typically quite small. In contrast, the long run marginal cost incurred to expand capacity can be quite large. In Tucson, Arizona, for example, initial water supplies were obtained from a central well field at a low marginal cost per gallon. When additional supplies were needed, the City purchased farmland in an adjacent valley for its water rights, thereby incurring a much higher marginal supply cost. Recently, when these supplies were insufficient, the federal government built the Central Arizona Project to import water hundreds of miles from the Colorado River. The cost to the city of this water is substantially higher even though the project is subsidized by the federal government. Thus the long run marginal cost of water to Tucson has risen with increasing quantities. Since the marginal cost of adding another customer to the urban network is usually quite small, the short run marginal cost and average cost are likely to be unchanged over a wide range of water use, only increasing when system capacity is reached. Therefore, in the long run the marginal cost and supply curve for water is likely to be in the form of stair-steps, being parallel to the quantity axis until a new source of water is needed (Kelso et al., 1973).

1.6 New Water Sources

When growth or other circumstances force the urban water provider into searching for new and more expensive sources, a number of options may exist. Perhaps, as in St. Petersburg, Florida, or Tucson, Arizona, it is necessary to bring distant well fields into production. Thus, the capital expense of new wells and pipelines must be incurred, as well as increased operating cost of pumping from greater depths or longer distances. In addition, it is likely that water rights must be purchased from farms, mines, or other holders before the water can be pumped. In some cases, it may be possible to trade treated effluent for the right to pump the best quality water from the well field or reduce the cost of the water from that field. The Los Angeles basin and a number of users in Nevada and Arizona purchase surface water from distant sources on the Colorado River (and Northern California in the case of Los Angeles). This practice necessitates regional use agreements, purchase from regional water markets, or negotiation with existing users to acquire the rights to the supply of surface water. In most of these cases, seasonal water supplied by snowmelt is the original source of the water. Thus, the timing of the water supply does not match well the summer seasonal peak use patterns of the consumers. While elaborate dams and canals have been constructed to handle this problem on a macro basis, individual water suppliers may also have the problem of where and how to store their water allocation until it is used (Agthe, 1986; Coe, 1980; Gleason, 1976).

In southern Arizona, the Los Angeles basin and other locations, surplus surface water is stored in previously depleted aquifers. This has the advantage of slowing or stopping subsidence caused by historic over-drafting of water by earlier water managers but raises a myriad of legal issues as the political, water district, and aquifer boundaries are unlikely to coincide. The potential for contamination of the stored water by overlying land uses outside the jurisdiction of the water company is frequently a threat as is pumping from the aquifer by another water user located over the aquifer. However, storage in a lake either near the source or near the market also has high costs in land acquisition and evaporation, particularly in desert areas. Surface water storage is also subject to potential contamination, and is more vulnerable to terrorists than aquifer storage. Distant water storage also poses potential costs associated with supply interruptions due to pump failure, earthquakes, or vandalism in a location where the water district has no political power to restrict land use.

1.7 Economic Analysis

The simplest economic analysis of the impact of prices on quantity decisions ignores all other possible influences on water use. However, a more sophisticated approach, used by serious analysts, employs econometric estimation of demand functions which include other related variables. Proper economic modeling requires identification of those variables that interact with the price and quantity variables or whose independent changes affect the price/quantity interaction (Friedman, 1967). Research on the effects of urban water price changes need to consider factors such as income, potential substitutes for water, climate, and special conditions surrounding water use such as government regulations. Personal income, for example, was found to be positively correlated with water use by Billings and Agthe (1980), Billings and Day (1989), Gottlieb (1963), Morgan (1973), Foster and Beattie (1979) and many other researchers.

1.8 Institutional Forms and Service Areas

Most urban water and sewer utilities are owned and operated by municipal government, although some are utility districts with their own elected boards of directors while others are privately owned and government regulated. Many states, like Arizona, have all three types of utilities. Water is not a pure public good since those who choose not to pay are excluded from its use and one person's use decreases what is left for everybody else. However, public health externalities, drinking water safety concerns, a feeling that equity in consumption of water across income groups is very important to society, economies of scale in provision, lack of private ownership rights to water itself, and huge costs of water development have resulted in publicly owned or publicly supervised urban water utilities. Boulding (1964) suggests that private ownership subject to strictly enforced safety rules and regulations is probably better than public ownership of water utilities as public ownership frequently leads to greater use of water by consumers than would be found under private ownership. This occurs because politically governed water utilities tend to keep water prices as low as possible, thereby subsidizing economically inefficient water uses.

Most public goods are financed based on either the benefits received principal (price) or the ability to pay principal (taxes). The first of these leads to the economic efficiency criterion that marginal cost to the consumer is equal to the value of the marginal benefit received by the consumer. The second of these involves equity across income levels and the assumption of positive social externalities being derived from allowing low income households to consume water at a lower price than higher income families. That is, if water is financed by property taxes, high income people with higher valued homes pay proportionately more for their water. Many public water districts are prevented by law from making a profit and most regulated privately owned water monopolies are regulated so as to earn no more than a reasonable rate of return on capital. However, neither of these strategies is likely to lead to an economically efficient water market.

We know of no public water provider that admits being a "cash cow" for general municipal operations. However, publicly owned electric utilities have been known to charge monopoly prices to raise money to offset (i.e. reduce) the contribution of local property taxes in communities dominated by transient populations associated with universities, military bases, or state capitols (Colberg, 1955). Also, some publicly owned utilities sell water to neighboring communities at a higher price per unit than to their own customers while escaping the scrutiny of the state public service commission because they are owned by a local political entity. On the other hand, in some communities people see publicly provided water as social good and provide a subsidy for the water provider from the general fund to the local water utility.

1.9 Scientific Choices and Public Choices

Victor Baker (1999) suggests that science has proven to be an effective force for human progress because of the pragmatic guidance that it provides for human action. Pollock (1998) and Portney (1992) have raised the issue that expert opinion is sometimes ignored by the voting public. Haerlin and Parr (1999) suggest that this problem may be due to the public perception of scientists working in "market based" science where they are required to justify the agenda of their employer. The public may also feel that science tends to force acceptance of solutions to problems rather than look at alternatives or consider value judgments. Baker (1999) points out that the realms of the scientist and the policy maker are "alien spheres" that create a gulf between science and policy with respect to water issues. However, science is valuable in drawing attention to and solving real world problems.

A good economic model (positive economics) should tell "what is" and be as free as possible from moral and ethical judgments. In this way, meaningful guidance can be offered for policy decisions concerning "what should be" (normative or policy economics). An outstanding example of this interrelationship is the use of linear programming models to solve economic resource use problems. The process usually "spits out" a single best solution to the problem of policy choice. However, frequently it is observed that several other policy solutions are very close in "pay-off" to the "best" solution arrived at. It may be that one of the "nearly best" solutions is more acceptable socially and politically. Since the data used in these analysis are often incomplete or of poor quality, the "nearly best" solutions may indeed have an equal or better "pay off" in reality. In these cases, the analyst should present all of the "nearly best" policy choices and leave the final decision to the elected or appointed decision makers. We present several economic models and discuss many policy choices in this book. The results from using these analyses and analytical tools should guide the final policy decisions on water issues, not determine them.

References

Agthe, D.E., "The Potential Costs and Benefits of Inground Storage of Imported Water," *Water Resources Bulletin*, 22 (1) February: 129-131.

Agthe, D.E., and R.B. Billings, 1997. "Water-Price Effect on Residential and Apartment Low Flow Fixtures," *Journal of Water Resources Planning and Management*, 122(1):20-23.

Baker, V., 1999. "Water Science and Policy in the Southwest," The Faculty Community Lecture Series," The University of Arizona, 4/28.

Billings, R.B., and D.E. Agthe, 1980. "Price Elasticities for Water: A Case of Increasing Block Rates," *Land Economics*, 56(1):73-84.

Billings, R.B., and M. Day, 1989. "Demand Management Factors in Residential Water Use: The Southern Arizona Experience," *Journal of the American Water Works Association*, 81(3):58-64.

Boulding, K., 1964. "The Economist and the Engineer," in S.C Smith and E.N. Castle, Eds. *Economics and Public Policy in Water Resource Development*, Iowa State University Press, Ames):82-92.

Coe, J.L., 1980. "Groundwater Storage for California Project," *Journal the Irrigation and Drainage Division* (ASCE):305-315.

Colberg, M. R., 1955. "Utility Profits: a Substitute for Property Taxes," *National Tax Journal*, 8, 382-387.

Foster, H., and B. Beattie, 1979. "Urban Residential Demand for Water in the United States," *Land Economics*, 55(1): 43-58.

Friedman, M., 1967. *Price Theory*, Aldine: The University of Chicago Press: 25.

Gleason, E., 1976. "Water Projects Go Underground," *Ecology Law Quarterly*, 5:625-668.

Gottlieb, M., 1963. "Urban Domestic Demand for Water: A Kansas Study," *Land Economics*, 39(2):204-210.

Haerlin, B., and D. Parr, 1999. " How to Restore Public Trust in Science," *Nature*, 400, August 5th: 499.

Howe, C.W., and F.P. Linaweaver, 1967. "The Impact of Price on Residential Water Demand and Its Relation to System Design and Price Structure," *Water Resources Research*, 3 (1):13-32.

Johnson, D., 1988. "Down the Drain," *The Los Angeles Times*, Feb. 2, Sect 5: 1.

Kelso, M.M., Martin, W.E., and Mack, L.E., 1973. *Water Supplies and Growth in an Arid Environment*, University of Arizona Press, Tucson.

Leftwich, R.H., 1966. *The Price System and Resource Allocation: Third Edition*, Holt, Rienhart, and Winston, Inc.:16.

Martin, W.E., and J.F. Thomas, 1986. "Policy Relevance in Studies of Urban Residential Water Demand," *Water Resources Research*, 22 (13):1735-1741.

Morgan, W.D., 1973. "Residential Water Demand: The Case from Microdata," *Water Resources Research*, 9 (4):1065-1067.

Morgenstern, R.D., Ed., 1997. *Economic Analysis at EPA*, Resources for the Future: Washington, D.C.

Pollock, R.A., 1998. "Imagined Risks and Cost-Benefit Analysis," *The American Economic Review*, 88 (2):376-380.

Portney, P.R., 1992. "Trouble in Happyville," *Journal of Policy Analysis and Management*, 11(1):131-132.

Stigler, G., 1966. *The Theory of Price*, The MacMillan Company, New York.

Wong, S.T., 1972. "A Model on Municipal Water Demand: A Case Study of Northeastern Illinois," *Land Economics*, 48 (1):34-44.

Chapter 2

Hydrology and Water Supply

Nathan Buras
University of Arizona

> **Water:** One of the simplest compounds in Nature: two hydrogen atoms combined with one oxygen atom. This minute, inaudible, odorless and texture free particle of no particular tang determined human destiny. Civilization evolved at the confluence of rivers. In all places, through all time, water has directed man's livelihood, success, economy and community. *As a natural force water is powerful.*[1]

2.1 The World Population and Water Resources

2.1.1 INTRODUCTION

During the Current Era (CE), the global population grew at an estimated rate of 0.04% per year until about 1650. Since then, the rate of growth rate increased somewhat more than geometrically, reaching an all-time high of 2.1% in 1956-1970 after which it has declined to about 1.6% per year. The world population today is approaching 6.5 billion and the United Nations projected a few years ago that in the mid-21st century it will reach 9.4 billion (Brown, Flavin and French, 1998).

There are significant differences in population growth among the various regions of the world: in the developing countries the population still grows at a rate of 1.9% per year, while in the developed regions the growth is of about 0.3-0.4%. In fact, about five

---

[1]Central Arizona Project, 2001

D.E. Agthe et al. (eds.), Managing Urban Water Supply, 11–22.

years ago there were a number of countries with zero, or even negative population growth, such as Belgium, Germany, Japan, and Spain. However, the world's *urban* population has increased since the mid-17th century by more than 3.5-fold, from about 0.74 billion to close to 3 billion. During this period, urban populations increased from 29% to over 45% of the total world population. In addition to the increased demand for imported food, much of which is grown with irrigation, the large migration of people from dispersed rural communities to urban areas of high density increased the urban need for water as well as the demand for wastewater treatment plants and reclamation facilities for the effluent so produced.

The world now has over 6.25 billion people. The population would double in a little over 40 years, if it continues to grow at the present rate of 1.6% per year. The amount of water on planet Earth at that time will remain the same as today, in many areas having a degraded quality that may make it unfit for human consumption. Consequently, there could be considerable social and political pressure to collect, treat, reclaim and reuse wastewater and, at the same time, to increase the number and efficiency of desalination plants.

There is a wide variation in per capita water use by country and region. In 2002, water use for all purposes in Tucson was 250 cubic meters per person per year, while San Diego used 230 m^3/person/year (Ganster, 2002). Considering only residential use, San Diego used about 153 cubic meters per capita per year during 1998, France used 106, Spain 94, Israel 65, Germany 46 and Greece 42 (World Resources Institute, 1998).

2.1.2 THE LAW OF THE MINIMUM

Since the planet Earth is finite in mass, volume, area and resources, we may pose the problem of water scarcity in a different way.

What is the size of human population that this planet can support?

The answer is a function of global requirements for various resources. Considering only two major requirements, food and water, the maximum size of the population can be expressed as:

$$\text{minimum}\begin{Bmatrix}\text{available water supply / individual water requirement}\\ \text{available food supply / individual food requirement}\end{Bmatrix}$$

This expression is an adaptation of the *law of the minimum* proposed about 150 years ago by the German agricultural chemist Justus Frieherr von Liebig (1803-1873). A more optimistic attitude is expressed by paraphrasing Antoine de Saint-Exupery,

our task is not to foresee the future, but to enable it.

2.1.3 THE DYNAMICS OF WATER ON PLANET EARTH.

Most natural phenomenon occurring on Earth can be traced to the energy generated by the sun, part of which continuously reaches our planet. Somewhat more than 70% of the surface of the Earth is taken up by oceans and other water bodies so that considerable quantities of vapor are generated by the energy reaching us from the sun. The water vapor ascends through the atmosphere to regions of lower temperature where clouds are formed. The water vapor condenses around tiny particles (mostly dust) and when reaching a certain size, gravity pulls them down to earth producing precipitation in the form of rain, snow and sleet. Some of the precipitation evaporates as it falls. The amount of precipitation is uniform neither in space nor in time. There are regions of the world, such as the Sahara and other deserts, where there is hardly any precipitation ever. Other regions are blessed with substantial amounts of precipitation, hundreds and even thousands of millimeters per year (1000 millimeters = 39.375 inches).

Precipitation does not occur uniformly throughout the yearly cycle. In a few parts of the world, such as the Northeastern Assam in India, it rains almost all the time. In most parts of the world, there are distinct wet and dry seasons. Southern Arizona is relatively fortunate since it has two rainy seasons, albeit with low amounts of precipitation: the monsoon season during July and August and the winter rains from December through February; total annual precipitation in the lower elevation areas averages less than 300 mm, while scattered mountain ranges receive 2-3 times as much.

As rainfall reaches the land, it is intercepted by vegetation: leaves of trees and shrubs or blades of grasses. Some of the rainfall so intercepted evaporates back into the atmosphere. The part which does not evaporate continues falling until it reaches the soil surface. Depending upon the microrelief, the rainfall water will accumulate in small depressions, or it will run off downgrade. In any event, as soon as the precipitation reaches the soil surface, some if it begins percolating below it. As soon as percolation begins, a major change occurs in the dynamics of the water delivered by rainfall. Water flows on the surface with a velocity of 0.6-1.0 m/sec. However, below the soil surface, water moves considerably slower – 3 to 6 orders of magnitude slower.

The annual water budget, averaged over all the land areas of the world, is precipitation = 800 mm, runoff (surface and subsurface) = 310 mm, evapotranspiration (evaporation and transpiration)[2] = 490 mm (Hornberger, Raffensperger, Wiberg and Eshleman, 1998). Observe that this budget is offered as volumes divided by areas, which is more convenient than just volumes. We are more familiar with a statement such as "Tucson airport received 5mm of rainfall", rather than "Tucson airport received 50,000 m^3 of water." Major water sources on planet Earth are given in Table 2.1.

[2] Transpiration is water vapor released through the leaves of plants.

Table 2.1 Major Stocks of Water on Earth.

	Volume (1000 km^3)	Percentage of Total Water	Percentage of Fresh Water
Total Water	**1,386,000**	**100**	
Saltwater stocks			
Oceans	1,338,000	96.54	
Salt/brackish groundwater	12,870	0.93	
Saltwater lakes	85	0.006	
Total saltwater	**1,350,955**	**97.476**	
Freshwater stocks			
Glaciers, permanent snow cover	24,064	1.74	68.70
Fresh groundwater	10,530	0.76	30.06
Ground ice, permafrost	300	0.022	0.86
Freshwater lakes	91	0.007	0.26
Soil moisture	16.5	0.001	0.05
Atmospheric water vapor	12.9	0.001	0.04
Marshes and wetlands[a]	11.5	0.001	0.03
Rivers	2.12	0.0002	0.006
Incorporated in biota	1.12	0.0001	0.003
Total fresh water	**35,029**	**2..524**	**100**

[a] Marshes, wetlands and water incorporated in biota are often mixed salt and fresh water. *Note:* Totals may not add due to rounding. Source: Shiklomanov, 1993.

2.2 Subsurface Waters

Groundwater is a very important source of water for urban uses. Table 2.1 indicates several categories of subsurface waters: *fresh groundwater* (about 30% of all the freshwater on planet Earth), *ground ice, permafrost* (less than 1% of all freshwater) and *soil moisture* (0.05% of all freshwater). Subsurface water is found in the pore spaces, voids, interstices and faults of various geological formations, including soils. When these pores, etc., are completely filled with liquid (saturated), the result is called *groundwater.* The upper limit of the saturated zone is called the *water table.* The formations above the water table are, of course, not water-saturated, forming the *vadose zone*. Groundwater is and always was an important component of drinking water supplied to urban areas as illustrated in Table 2.2.

Table 2.2 Groundwater as a Share of Drinking Water Use, by Region

Region	Percent of Drinking Water from Groundwater	Number of People Served by Groundwater (millions)
Asia Pacific	32	1,000 – 1,200
Europe	75	200 – 500
Latin America	29	150
United States	51	135
Australia	15	3
World		1,500 – 2,000

Data from late 1990s: Data on Africa not available. Source: Brown, Flavin and French, 2001.

2.3 Mining of Groundwater

It appears that there is a universal tendency to extract water from some aquifers, on a yearly basis, at rates exceeding those of their natural replenishment (recharge). The mining of aquifers can result in a number of undesirable outcomes, such as lowering of water tables thus requiring more energy for pumping water from greater depths; decreasing (and sometimes preventing) seasonal flows in streams hydrologically connected with over-pumped aquifers; inducing seawater intrusion into coastal aquifers and land subsidence.

In the United States, the largest overexploited aquifer is the Ogallala, which underlies much of the central part of the country; 450,000 km^2 of land stretching from South Dakota to the Texas panhandle. It varies in thickness from 330 m to 30 cm, with an average of 60 m. The aquifer is not an underground lake or river, but is more akin to a gigantic sponge composed primarily of saturated sand and gravel. The average annual recharge of this aquifer is estimated at 6-8 km^3/year while its average annual use is about 22.2 km^3/year, mostly for irrigation (Gleick, 2000). From the 1940s to 1980 the water table of this aquifer fell an average of three meters, with some areas of Texas incurring a decline of 30 m or more.

Concern for the future of this water supply has led to various efforts to improve water use efficiency. For example, scientists at Texas A&M University and the state and federal governments have set up the North Plains Evapotranspiration Network which monitors weather and crop conditions and advises farmers, on a real time basis, of the optimum timing and duration of irrigation to maximize yields and minimize water use. Farmers subscribing to one of the information alert networks saved an estimated 76 million m^3 of water with a pumping cost of $5 million in one year without reducing crop yields. At least one school district is using this service to determine its turf irrigation scheduling, and the developers of this program hope to extend its application to urban water users. As a result of these and other conservation efforts, the water table fell an average of only 30 cm between 1980 and 1990 (Cormis, 2000). Additionally, the Texas legislature recently authorized the establishment of 80 local groundwater conservation districts to regulate groundwater extraction and exportation. This was in response to extensive plans by entrepreneurs to mine groundwater in the Texas panhandle and export it to cities hundreds of kilometers away.

On the other side of the continental divide, large parts of urban Arizona also rely on overused aquifers. The state legislature finally reacted to this ongoing problem by enacting the *Arizona Ground Water Act of 1980.* This legislation identified four regions in the State where the aquifers were heavily exploited: Tucson, Pinal, Phoenix and Prescott. These regions were designated as *Active Management Areas* (AMAs).[3] The function of the AMAs is to manage their respective water resources such that by the year 2025 they will reach equilibrium between the amount of water pumped from their aquifers and that recharged into them, naturally and/or artificially.

In the Tucson area, the heavy exploitation of aquifers from the early 1940s resulted in the drying out of the Santa Cruz River. However, the big storm of October 1983 generated a very large streamflow in the Santa Cruz River and its tributaries, causing significant damage because buildings and infrastructure had been developed within its floodplane. It was estimated that the Santa Cruz in Tucson, for a few hours, had a discharge of about 1,000 m^3/second. An additional result of this event was an upward revision in the definition of a 100 year flood.

Over-pumping of coastal aquifers, for example in Southern California and Israel, induced the intrusion of seawater, thus taking out of commission a number of productive wells. If no counter-measures are taken, the seawater intrusion will advance inland and more well fields could be abandoned. To remedy this highly undesirable situation, a fresh-water barrier can be established a short distance inland from the coastline and parallel to it, by drilling a number of wells of appropriate depth. Fresh water is injected into these wells at a rate determined by the physical characteristic of the aquifer and the distance between the wells. The major challenge for this plan is obtaining additional sources of fresh water to inject.

Excessive exploitation of aquifers over extended periods of time often results in subsidence of the land. As the water table is continuously lowered, the weight of the dry material above it collapses it, causing the land surface to subside. In Tucson, aquifer compaction and associated land subsidence of about 15 cm has accompanied the long-term water level declines (Arizona Department of water Resources, 1994). Chapter 12 discusses land subsidence in Houston, Texas and Chapters 13 and 14 also deal with this problem.

2.4 Storage of Surface Water in Aquifers, Dams and Reservoirs

The Amazon River in South America carries about 15% of all the world's streamflow. In order to utilize such flows, large capital investments in diversion, storage, and transmission are often required. One of the longest aqueducts is the Central Arizona Project which is 540 km long and cost billions of dollars (Wilson, 1999). Even in Brazil,

---

[3]Today there is a fifth AMA for Santa Cruz County which adjoins Mexico.

southeast of the mouth of the Amazon River, there is a region *(poligono da seca)* which has scarce water resources and could use some of the Amazon water, yet no such project was designed. There are serious limitations for transferring surface water long distances because of the required resources and political constraints. Until relatively recently, the environmental impacts of water diversion were generally ignored. It is now widely recognized that diversions have adverse impacts on aquatic populations, including numerous commercially valuable species.

One of the major discrepancies between the estimated regional amount of water resources and their desired availability has the dimension of time. Most surface streams and rivers exhibit a discharge variable in time, depending primarily on the precipitation regime.[4] The demand for water from a stream or river has a time pattern different from that of the surface hydrologic phenomena. To overcome this discrepancy, reservoirs are built to store water when it appears in quantities exceeding current requirements, making it available for use during periods when natural flows are insufficient to meet demands. One common method of building such reservoirs is to construct dams at appropriate sites in a river basin.

In some locations such as southern California and Arizona, river water is also stored in carefully selected aquifers. Water pumped from the Colorado River or available from local rivers on a seasonal basis is recharged into the aquifer, either through settling basins or by injection wells. The major advantages of aquifer storage are that water is not lost to evaporation once it is stored and the avoidance of building reservoirs with their high costs and environmental impacts. Another advantage of aquifer storage is that it raises the water table, potentially reducing future subsidence. A major challenge facing aquifer storage is maintaining control of the stored water. A clear understanding of the hydrology of the aquifer is necessary to avoid loss of the stored water to subsurface flows out of the aquifer and pumping of the stored water by unauthorized well owners. Clear and enforceable ownerships rights to the stored water are essential.

Reservoirs are operated by releasing water stored behind a dam. Reservoir operation is a decision process, i.e., an alternative is selected from among the options available in the range from zero releases to actually emptying the reservoir. These decisions are made at discrete points in time.

Water is released from reservoirs in order to attain one or more objectives: meeting municipal (domestic and industrial) demands, supplying water for irrigation and other requirements of agricultural production, generating hydropower, creating a storage volume in the reservoir capable of accommodating high streamflows (floods) and others. In fact, when operating a reservoir, there are two mutually exclusive, yet inter-related, sets of decisions:

- How much water to release during each time interval
- How much water to leave in the reservoir.

---

[4]One exception is the Dan River in Israel. Dan is one of the three major sources of the Jordan River. Its almost steady discharge is due primarily to the specific geohydrological environment.

Yet there is a basic equation representing this apparent paradox regarding reservoir operation: the *Law of Conservation of Matter:*

> The amount of water stored at the beginning of the next period of time
> = The amount of water stored at the beginning of the current period
> + inflow to the reservoir during the current period
> + precipitation falling on the reservoir during this period
> – evaporation losses from the reservoir
> – seepage losses from the reservoir
> – uncontrolled spills from the reservoir
> – intentional releases of water from the reservoir.

Because the decision of how much water to release is dependent on the decision of how much water to leave in the reservoir, the problem of operating a surface reservoir is indeterminate. The decisions related to the operation of a surface reservoir are made considering their possible outcomes. Positive outcomes are known as ***benefits*** while negative outcomes are called *costs*. Benefits and costs are usually expressed in monetary terms, which are rather difficult to determine accurately, or even satisfactorily.

An alternative way to express the outcomes of decisions regarding the operation of surface reservoirs is to use volumes of water. Then water releases and water storage targets are specified in connection with the different objectives of the system. The outcomes of the reservoir operating decisions indicate whether the targets were attained or not. These outcomes are dimensionless and if a target is completely attained, its value is unity. If a target is not attained at all, its value is zero. These functions are often called utilities; they may indicate whether targets are only partially attained and they are not necessarily linear.

2.5 Conjunctive Use of Surface and Subsurface Waters

Surface streams and subsurface waters are components of a single continuum, as indicated by the fact that excessive exploitation of aquifers can and often does affect negatively flows in adjacent streams. This indicates that the full development of water resources in a region often requires the operation of surface reservoirs in conjunction with ground water aquifers. The aquifer, then is not only a source of water, it may also store part of the stream flow reaching the surface reservoir. This system requires an operating policy that specifies the releases from the surface reservoir, the amounts of water extracted from the aquifer and quantities of water to be recharged into it (Buras, 1963).

In conjunctive use projects, the surface water source and the aquifers need not be in close geographical proximity. A rather extreme example is the "Clear Water" Project in Tucson, Arizona. Colorado River water is conveyed to the Tucson area by means of an aqueduct almost 540 km long (Tucson Water, 2002). The Colorado River water is recharged into aquifers through spreading basins, it "blends" with the groundwater in the underlying aquifers, the "blend" is pumped out through production wells and delivered to urban users in the Tucson Metropolitan Area.

The success of the conjunctive use strategy depends on the interaction of a number of important variables: the quality -- chemical and microbiological -- of the surface water and groundwater; pricing of water and of energy (used primarily for pumping groundwater) and the cooperation of the water users. Considering the chemical water quality in the "Clear Water" Project, Tucson groundwater has a mineral content of about 300-350 parts per million (ppm), while Colorado River water has a concentration of total dissolved solids (TDS) of 780-790 ppm. It is clear that the blended water supplied to the urban users in Tucson will contain more than 300-350 ppm TDS and this concentration will creep upwards in time. Although the Tucson Water agency claims that the concentration of TDS in the "blend" will not exceed 450 ppm in the foreseeable future (Tucson Water, 2002), it is obvious that in the more distant future, the concentration of TDS will increase beyond that limit.

It is also obvious that the "Clear Water" project is a capital intensive proposition: its purpose is limited to only one objective -- urban water supply -- delivered by an aqueduct about 540 km long and lifting water 805 m. One might conjecture that the Colorado River water could be used for agricultural activities in the Avra Valley west and south of Tucson near the end of the canal system and by farmers along the canal northwest of Tucson. This, however, is not very likely to happen under current institutional arrangements, considering that these farmers prefer to irrigate their crops with groundwater pumped with subsidized electricity, rather than purchasing the more expensive Colorado River water delivered by the Central Arizona Project. This problem of inappropriate relative prices for water of different qualities is discussed in Chapter 13.

2.6 Water Quality Issues

The quality of our waters is often considered to be a national problem; however, it is more likely to be a bundle of local and regional problems (Dorfman, Jacoby and Thomas, 1972). The quality of waters has several distinct dimensions: chemical, physical, microbiological, thermodynamic and others. These dimensions vary not only in space (locally, regionally) but also in time (seasonally and annually). Much of the variation is anthropogenic (i.e. caused by humans).

The *chemical* quality of water is usually expressed in terms of *total dissolved solids*, TDS. This is only a rough quantification of the chemical quality of water. As mentioned in Section 2.5, Tucson groundwater supplied for domestic use has a TDS concentration of 300-350 ppm. Potable water will be quite acceptable at TDS concentrations of 500 ppm and even somewhat higher. For irrigating crops, the situation is different: some crops are quite sensitive to waters having a TDS of 500 ppm, while other crops are considerably more tolerant to higher TDS concentrations.

Of great importance is the nature of the ions in solution. The concentration of dissolved oxygen is of major importance, since oxygen dissolved in water is a crucial factor in maintaining the flora and fauna of water courses and lakes. However, oxygen is only slightly soluble in water and its solubility decreases with increased water temperature. Its highest solubility is at 0^0 C -- 14.6 mg/l. At 25^0 C it is only 8.4 mg/l (Thomann and Mueller, 1987). Discharging wastewater into water bodies introduces substantial quantities of organic material that are decomposed by microorganisms which use dissolved oxygen (DO) in this process. The result may be an anaerobic state of the water body, which is highly undesirable. For this reason, Harold Streeter and Earle Phelps studied the concentration of DO in the Ohio River at the beginning of the 20th century and developed the sag curve which describes the DO deficit variation with distance in a river (Thomann and Mueller, 1987).

Another instance illustrating the importance of the composition of the TDS is the unsuccessful attempt to introduce Colorado River water directly into a branch of the Tucson potable water distribution system in 1992. The Colorado River water supplied by the CAP had a much higher concentration of TDS than the groundwater that was for a long time the only source of water. Much of the matter in solution in the Colorado River water consisted of calcium sulfate (gypsum) while the distribution pipe network had internal deposits of other minerals. When the CAP water was released into the distribution pipes, a chemical reaction occurred between the matter dissolved in the Colorado River water and the pipe coatings. The result was a foul smelling suspension of a dirty brown color, which was not only un-drinkable but also damaged some appliances. The Water Department of the City of Tucson had to pay substantial damages to the affected public and the citizens of Tucson approved a proposition forbidding the City to directly deliver CAP water for domestic use.

2.7 Wastewater Treatment, Reuse and Disposal

Wastewater is the spent or used water of a community, comprising water-carried wastes from residences, institutions, commercial buildings and industries. *Sewage* is the liquid waste of a community conveyed by a sewer. Thus both terms have the same meaning. In recent usage the word "wastewater" has taken precedence (Gupta, 2001).

The quantity of wastewater (sewage) generated by a community depends on its population, its climate and people's life style. In general, it is estimated that about 60-80% of the total water supplied to an urban community becomes wastewater. As a result, wastewater may be considered as a source of water, if treated appropriately, or as a municipal burden because it needs treatment and disposal.

The purpose of municipal wastewater treatment is to prevent contamination of the water bodies into which the treated effluent is discharged. Wastewater treatment consists of a series of processes of varying complexity, usually activated in series, leading to improved quality of the treated effluent (Viessman and Hammer, 1998). Wastewater can be treated to the extent that the effluent is fit to be supplied as drinking water, albeit this being a rather costly proposition. For example, much of the drinking water supplied to Windhoek, the capitol city of Namibia (southwest Africa), originates from treated wastewater.

In almost all cases, the treatment of municipal wastewater is quite short of drinking water standards. Most common is a combination of primary and secondary treatment which removes about 50% of suspended solids while the remaining organic matter is extracted with biological processes. However, the resulting effluent is still rich in TDS, contains some suspended solids, organic matter and excreted pathogens. The re-use of treated municipal wastewater, for whatever purpose, involves significant public health risks.

Treated effluent is disposed in rivers, lakes, the ocean, or is used for irrigating crops. One major disposal system into the ocean is at the Hyperion wastewater treatment plant in Los Angeles. The treated effluent is discharged into the Pacific Ocean through a seven mile (11km) pipe line ending on the ocean floor. Occasionally, ocean tides bring some of the effluent, which is not completely treated, to adjacent beaches.

2.8 Increasing Costs of New Sources of Municipal Water

A prominent feature of water resources systems all over the world is increasing urbanization. Populations of existing towns and cities continue to grow, while rural villages and hamlets which historically had more or less constant populations are experiencing rapid growth. These demographic phenomena result in new patterns of municipal water demand and have an influence on the quality of the environment.

Several aspects are relevant, of which two will be mentioned. One aspect is that an augmentation of water supply to a city may necessitate the tapping of sources located at greater and greater distances. The Central Arizona Project was presented above in some detail and it is emphasized again that the delivery of Colorado River water to the Phoenix-Tucson region overcomes a topographical obstacle of over 800 m. This requires substantial amounts of energy, so that the CAP water is expensive and the cost of its delivery is likely to increase in the future because of the increasing cost of energy.

The other aspect relates to the wastewater generated in urban areas. Depending on the degree of treatment of the wastewater and the disposal of the treated effluent, the sustainability of the environment may be threatened and downstream communities and activities may be negatively affected. If and when these negative effects are compensated financially, the cost of supplying the water to the urban area will increase.

In Tucson, as in many other locations, groundwater could be pumped from greater depth. To do so, at least three cost factors need to be considered:

- Additional amount of required energy
- Higher concentration of TDS
- Increased risk of land subsidence

Most of the analytical thinking related to the supply and management of water in the urban environment makes use of models. Many models represent only a fraction -- often only a very small fraction -- of the environment, so that the actual situation, in all its complexity, is not in the model. Continual improvements in models are occurring, however. For example, water flows in the San Pedro River basin were recently modeled, and a prediction was made that stopping pumping on a selected six mile (9.7 km) reach of this small Arizona river would restore perennial surface flows. The Nature

Conservancy subsequently bought the ranch which contained this stretch of river, stopped the previous year-round irrigation of alfalfa, and was rewarded by the predicted restoration of surface flows.

Since water -- the hydrosphere -- is a major component of the environment, decisions related to the supply and management of water for an urban area have a significant effect on the local and regional environment. Current supply and management decisions will almost certainly affect the availability of these resources for future generations. These two important points -- the current quality of the environment and the effect of current water using activities on the world that future generations will inherit -- are not yet satisfactorily resolved. Hopefully, understanding, wisdom and imagination will help to resolve these issues (Buras, 2001).

References

Arizona Department of Water Resources, 1994. *Arizona Water Resources Assessment, Volume II, Hydrologic Summary,* Phoenix, Arizona.

Brown, L.R., C. Flavin and H. French, 1998. *State of the World 1998*, Norton, New York.

Buras, N., 2001. "Water Resources -- Unresolved Issues," An Editorial, *Journal of Water Resources Planning and Management*, 127(6): 353.

Buras, N., 1963. "Conjunctive Operation of Dams and Aquifers," *Journal of the Hydraulics Division*, Proceedings of the American Society of Civil Engineers, 89(HY6): 111-131.

Central Arizona Project, 2001. *Annual Rreport,* Phoenix, Arizona.

Cormis, D., 2000. *Saving North America's Greatest Aquifer,* www.nps.ars.usda.gov.

Hornberger, G.M., J.P. Raffensperger, P.L. Wiberg & K.N. Eshleman, 1998. *Elements of Physical Hydrology,* Johns Hopkins, Baltimore, Maryland.

Dorfman, R., H.D. Jacoby and H.A. Thomas, Jr., editors, 1972. *Models for Managing Regional Water Quality,* Harvard University Press, Cambridge, Massachusetts.

Ganster, P., Ed., 2002. *The U.S.-Mexican Border Environment,* SCERP Monograph Series no. 3, San Diego State University Press, San Diego, California.

Gleick, P.H., 2000. *The World Water 2000-2001*, Island Press, Washington, D.C.

Gupta, R.S., 2001. *Hydrology and Hydraulic Systems*, Waveland Press, Inc., Prospect Heights, Illinois.

Shiklomanov, I.A., 1993. "World fresh water resources," in *Water in Crisis*, Peter H. Gleick, editor, Oxford University Press, Oxford, U.K.

Tucson Water, 2002. *2001 Annual Report*, City of Tucson, Tucson, Arizona.

Thomann, R.V. and J.A.Mueller 1987. *Principles of Surface Water Quality Modeling and Control*, Harper and Row, New York.

Viessman, W., Jr. and M.J. Hammer, 1998. *Water Supply and Pollution Control* (sixth edition), Addison-Wesley, Menlo Park, California.

Wilson, D.S., Jr., 1999. *Central Arizona Project, Annual Report*, Phoenix, Arizona.

World Resources Institute, 1998. *World Resources: A Guide to the Global Environment, Environment Change and Human Health*, Oxford University Press, New York.

Chapter 3

Price Rationing

Donald E. Agthe
Economic Consultant
R. Bruce Billings
University of Arizona

3.1 Water, a Scarce Good

A *scarce good* can only be obtained by giving up something else or by paying a positive price for it. In most urban areas, water is a scarce good and is therefore subject to the laws and principles of economics. In particular, the quantity of water taken from the market by any given buyer or group of buyers is influenced by the price that must be paid. The price of water, in its most general sense, includes not only monetary payments, but the time and energy expended to obtain that water. For example, if a person must drive or walk some distance to buy water, the price includes the value of time spent as well as the monetary payment. As urban water becomes increasingly scarce, price can be used to allocate its use efficiently among its many competing end uses. Prices can also be used to encourage the meeting of community social goals with regard to the availability and use of water.

3.2 Demand

Demand is defined as a schedule of the quantities of a good or service that a specific group of consumers is willing and able to take from the market at each of many possible prices within a given time period. Note that there is an 's' on both price and quantity. A

D.E. Agthe et al. (eds.), Managing Urban Water Supply, 23–43.

price increase generally results in a move to a smaller quantity unless the quantity cannot be reduced for technological reasons. Similarly, a price reduction often leads to higher use levels unless legal restrictions limit water use. Demand is not the same as needs or wants. Not only do people have to desire the product but they also must be able to pay for it. Many people would like to have green lawns, even in a desert environment, but will do so only if they can afford it. Similarly, people might like to take long hot showers, but will do so only if the cost to them of heating and delivering water to their shower is low enough compared to their income. The ability to pay problem raises the question, to be discussed later, of what to do about pricing water for low income families who can't afford a green lawn and have difficulty affording water for necessary uses such as washing, cooking, and drinking.

The definition of demand includes the concepts of market, good, and time period. Water use may be measured over time periods ranging from less than an hour to a year or more. Defining the time period of the demand relationship is important because annual demand as well as peak daily and seasonal demands influence water system design, water purchase contracts, reservoir and pumping operations, pricing policies and conservation programs.

In any water demand analysis, the quality of water as a good must be defined. For example, the good may be raw river water, potable water or reclaimed wastewater. Many uses do not require drinking quality water. Many industrial and recreational demands could be satisfied by lower quality water such as reclaimed wastewater or untreated river water, as long as it meets a specified minimum standard. Defining the scope of the market is also necessary. Geographically, one could analyze water demand in a basin shared by several water providers, the demand faced by a single water provider, or the national or world wide demand for water.

Many water providers divide their customers into sub-markets such as household, industrial, recreational, and governmental as the demands in each can differ. Economic development policies and other considerations may dictate a different pricing structure for each market. For example, low priced water for industry may help provide jobs for the community. When the water provider is capable of delivering water of different quality levels, the demand for each type of water should be analyzed separately. For example, drinking quality water should have the highest price reflecting its higher cost of production and value while reclaimed wastewater should carry a lower price reflecting its lower value in many uses.

3.2.1 LAW OF DEMAND

The *law of demand* states that as the price of an item increases the quantity taken from the market decreases or if the price of an item decreases more will be taken from the market. This law is the heart of price rationing. Since water is an economic good, the quantity of water purchased will generally fall when the price of water is raised. This observation is supported by numerous empirical studies. Therefore, price changes can be used to "control" the quantity of water demanded and price rationing can be an important tool for water utility administrators. Even though large price increases are likely to be effective in reducing the quantity of water demanded, they may be politically unacceptable. Thus, other measures are often combined with price rationing to achieve water conservation goals.

A simple linear demand schedule is presented in Table 3.1. For each given price there is a unique quantity, and since this is a linear curve intermediate points can be readily interpolated between those given. Note that the demand schedule includes many price-quantity combinations pertaining to a specified group of water users during a given time period.

Table 3.1 Linear Demand Schedule

Price per Unit	Units Bought per Month
$5.00	0
4.50	1
4.00	2
3.50	3
3.00	4
2.50	5
2.00	6
1.50	7
1.00	8
0.50	9
0	10

The same information is presented as a demand curve with price on the vertical axis and quantity on the horizontal axis in Figure 3.1. The price and quantity are inversely related, demonstrating the law of demand. Plotting price on the vertical axis is a convention adopted by British and North American economists. Quantity is actually dependent on price. The simple demand schedule in Table 3.1 and demand curve of Figure 3.1 could represent the demand for an individual water customer, a class of customers, or an entire city or region. The demand curve shown is linear only for illustration purposes. Real world demand curves are non-linear.

Although urban water is usually provided by a public entity or a government regulated private entity, it is not a pure public good since consumption by one customer precludes consumption of the same water by another customer.[5] Thus, demand curves can be added horizontally. For example, suppose that at a price of $1.00 per unit, Jim, Sam, and Mary would buy 1.3, 1.0, and 0.7 units per month respectively. Their total demand at a price of $1.00 would be 3.0 units of water. However, at a price of $.50 they would buy 2.8, 2.2 and 1.7 units respectively. Thus a second point on their aggregate demand curve is 6.7 units per month at $.50, as shown in Figure 3.2.

[5] Urban water supply is actually a quasi-public good based on the existence of public health externalities and economies of size in distribution system construction.

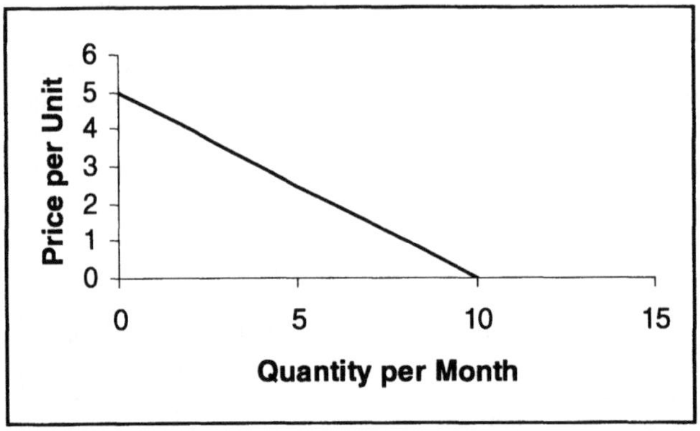

Figure 3.1 Linear Demand Curve

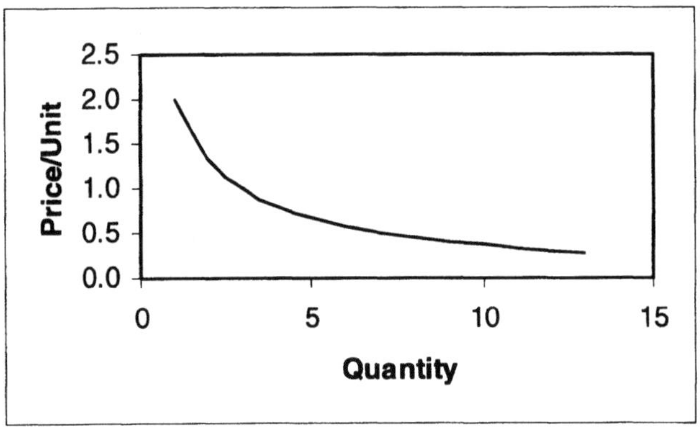

Figure 3.2 Typical Demand Curve

Additional points on the demand curve would show the amount these three people would buy at each possible price. If the only data available were the two price-quantity combinations discussed above, the analyst would be able to interpolate other points on the demand curve between these points. However, these two points alone do not provide much insight into the effect of prices dramatically lower than $0.50 or higher than $1.00.

3.2.2 DETERMINANTS OF DEMAND

The demand curve gets its shape and location because of the effect of price and several non-price determinants of demand, including:

- Tastes and preferences
- Price of substitutes
- Number and closeness of substitutes
- Income
- Price of complementary goods and services
- Climate (rainfall and temperature)
- Population
- Number of actual and potential uses

People have varying tastes and preference for water-using goods such as lush gardens, swimming pools, multiple bathrooms, and lengthy showers. Some people prefer drinking bottled water instead of tap water, even in areas with high quality water systems. Most water use by households, business and government has substitutes. Leak check programs, plumbing repairs, recycling systems for industrial cooling, and low flow toilets and faucets are all labor or capital substitutes for water. A change in the price of water relative to these substitutes will encourage or discourage their adoption.

The prices of complementary goods, those used with water, also influence water demand. Assuming that new swimming pools replace *non*-irrigated areas, fewer pools will be built and less water will be used in the future if swimming pool construction costs rise. Water use increases with income as people add features such as swimming pools, lawn sprinkling systems, multiple bathrooms, and decorative fountains. Water use generally increases in direct proportion to population, so utility planners need to be responsive to population projections. Finally, urban water demand is generally larger in the summer because of landscape and garden irrigation and the use of water for cooling.

The economic quantity of water depends upon its quality. High quality water with more uses tends to have a greater economic value and demand than water with fewer uses. For example, potable water tends to command the highest price because all customers of the water provider can use it while reclaimed water, which is of lower quality with fewer economic uses, commands a lower price.

3.2.3 CHANGE IN DEMAND AND CHANGE IN QUANTITY DEMANDED

A change in *demand* is a shift of the entire demand curve, as shown in Figure 3.3, due to a change in a non-price determinant of water use. On the new demand curve to the left of the original curve shown in Figure 3.2, the quantity of water demanded at $1.00 is 6 units compared to only 3 units on the original curve. Such a shift might occur when homeowners decide to put in larger irrigated lawns, and they are allowed to buy as much water as they want at a constant price. In contrast, a change in the *quantity demanded* of water only results from a change in the price of water. This was shown in Figure 3.2, where the quantity of water purchased is larger at $0.50 than at $1.00, but the demand curve has remained constant. As one of Agthe's astute students correctly pointed out, the quantity taken from the market increases in both of these examples. However, it is how these changes take place that is crucial for urban water use policies.

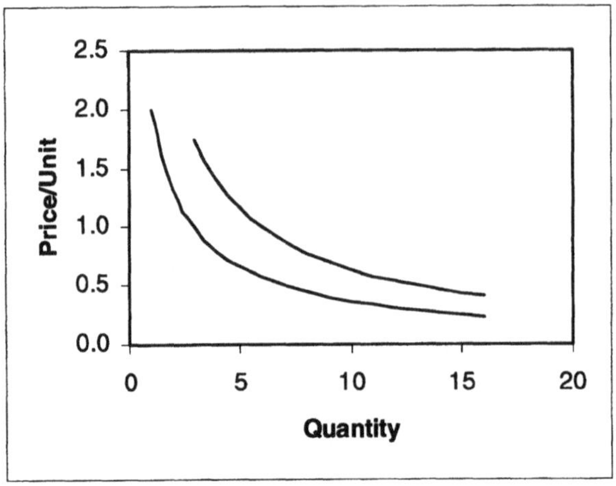

Figure 3.3 Increase in Demand

Raising price to reduce the quantity demanded, thereby achieving conservation, is an effective policy instrument. However, increasing the price of water has both a *substitution effect* -- less water is consumed -- and an *income effect* -- consumers are less able to consume other goods because their water bills are larger. If the consumers who are voters react negatively to the price increases and reject them as being too harsh, a very useful policy instrument might be lost in the political arena.

A decrease in demand occurs when less water is sold at the same price or the same quantity is sold for a lower price. A lower price accompanied by a lower quantity is a decrease in demand. Conversely, an increase in both price and quantity taken is an increase in demand. Some examples of an increase in demand are:

- Population increases
- New industrial or recreational customers locate in the area
- Increasing incomes allow consumers to purchase more water using goods such as swimming pools, bathrooms, water intensive landscaping, and decorative fountains
- The price of swimming pool construction falls
- Very low income residents connect to the water system for the first time as they obtain higher incomes or government subsidies

In addition to raising price, policy responses in water short areas with increasing demand for water could include requiring less water intensive landscaping, slowing population growth, discouraging water intensive industries, and making pool construction permits more difficult to obtain.

3.2.4 DERIVED DEMAND

Johnson (1988) found that consumptive uses such as bathing, hygiene, washing, cooking, and drinking, constituted only 12 percent of the water demand in Los Angeles. Most of the other demand was for water used to produce goods such as industrial products, clean cars, recreation, and attractive landscapes. These demands are known as *derived demands* because they are derived from the demand for the water using product and/or the water demand arising because of the production function of the process used to produce the good. The existence of derived demand implies that some non-price policies to reduce water demand could be aimed at altering the production functions of these goods. For example, the size of green areas in golf courses could be reduced or an industrial plant could be required to recycle its cooling water.

3.3 Supply

Supply is defined as a schedule of the quantities of a good or service that producers are willing and able to offer in the market at all possible prices in a given time period. The price-quantity combinations on the schedule are dependent upon the willingness and ability of the producers to provide the good at each price. As with demand curves, time period and careful definitions of market and the good are important to any analysis. The law of supply states that price increases generally lead to greater quantities of the good being brought to market.

While this market-oriented definition of supply might work well in regional water markets where water rights are traded by multiple producers and buyers, most urban water markets are characterized by monopolistic public local water utilities or regulated private providers. Many public water providers are limited by law to recovering their cost of production and regulators impose a fair rate of return restriction on private water companies. These producers find themselves in the position of having to adjust supply to meet demand at a nonprofit or low profit price usually determined by state or local officials. They may also face political constraints on their ability to raise rates to finance new capital expenditures. Thus, most of our discussion of supply will not follow the free market model.

3.3.1 DETERMINANTS OF SUPPLY

The most important non-price determinants of the shape and location of the supply curve are:

- Technology of production, treatment, and delivery
- Prices of inputs used in production
- Number of actual and potential sellers
- Governmental regulations

Technology, represented by the production function for water, plays an important role in determining the supply curve. Costs of production vary dramatically by raw water source between surface water, ground water, and desalinized ocean water, as discussed in Chapter 4. Similarly, methods of transportation, drilling, pumping and treatment affect the supply curve of urban water. Water production inputs include labor, land, capital, and energy. These inputs are all necessary, so are complementary in production. However, when alternative sources of water and technologies of transmission and treatment are considered, they are also substitutes to some extent. For example, meter reading can be done in the traditional way by having someone go to each meter and record the information or by remote telemetry. Therefore, the absolute and relative prices of inputs to the production of water are important supply curve determinants.

3.3.2 THE SUPPLY CURVE

Since technology, sources of water, and other policies vary among urban water providers, supply curves can be found in various shapes. Generalizations of those most common to water are shown in Figure 3.4. Figure 3.4a depicts a typical supply curve illustrating the law of supply that states that as price increases, quantity offered increases. Figure 3.4b shows a perfectly inelastic supply curve in which output is at the maximum possible rate of production so that higher prices do not evoke higher output. This is generally a short run situation in which no new water sources or capital investments are considered. Figure 3.4c is a good general representation of urban water supply, combining elements of figures 3.4a and 3.4b. This shows output increasing in response to higher prices up to a maximum quantity of water above which higher prices do not elicit any additional output. This happens when the urban water provider is operating at maximum capacity. An *increase in supply* is shown by a rightward shift of the supply curve caused by a change in a non-price determinant. In contrast, an *increase in quantity supplied* is shown by a move along the curve to a larger quantity caused by a change in price. Generally speaking, supply has increased whenever the same quantity is offered at a lower price or a higher quantity is offered at the same price.

3.3.3 LONG RUN VS. SHORT RUN SUPPLY

The supply curves depicted in Figure 3.4 are typically short run. When the water provider experiences a perfectly inelastic short run supply curve and increases in demand, it may choose to seek new sources of water which are likely to be more expensive. On a desert island, desalinization of seawater might be the best alternative. In other regions, surface water from long distances could be imported. Tucson, Arizona uses Central Arizona Project water from the Colorado River which is over 200 miles (322 km) away. Some cities have acquired well fields in neighboring regions. For example, St. Petersburg, Florida replaced local well fields subject to salt-water intrusion with wells outside her area. In larger metropolitan areas such as San Diego California, El Paso Texas, and Tucson, it may be efficient to produce and market reclaimed water to those capable of using it. Regardless of specific sources, the long run supply curve

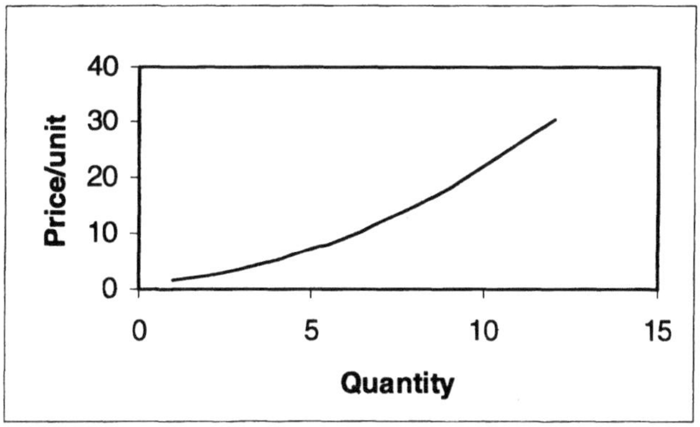

Figure 3.4a Traditional Supply Curve

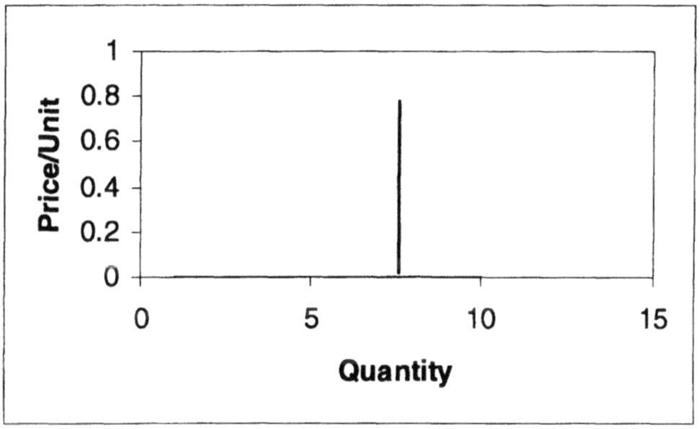

Figure 3.4b Perfectly Inelastic Supply

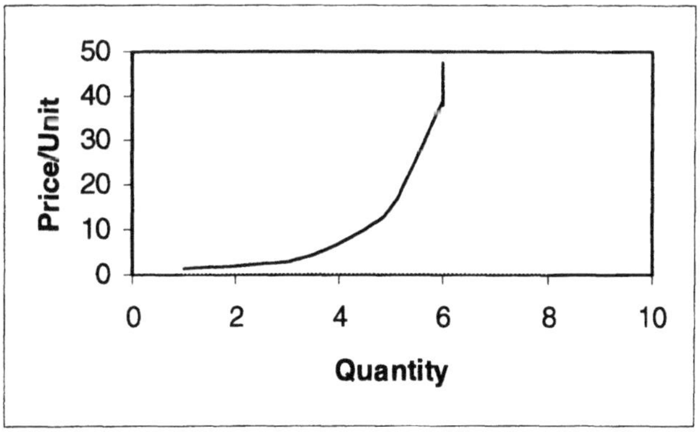

Figure 3.4c Typical Urban Water Supply Curve

would look like Figure 3.5 with each step representing the next higher cost source of water.

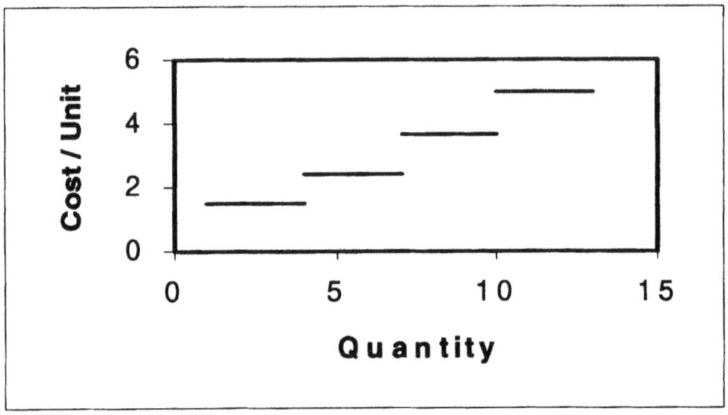

Figure 3.5 Long Run Supply Curve

3.4 Internal and External Cost

The total cost to society of supplying water can be divided into internal and external categories. Outlays for capital, labor, and energy are *internal* costs, paid by the water utility. *External* costs result from subsidence or ecological damage due to over pumping of the aquifer or fish deaths because of a dam to provide surface water. External costs are not costs to the water provider although they represent costs to society of providing the water. Thus, a water provider restricted to pricing its water subject to cost of production may only include the internal costs in its calculations. Community leaders who benefit from cheap water policies would be likely to support such a position while downplaying or ignoring environmental concerns.

Utility decision makers often ignore damages from over drafting and the return of wastewater in less than pristine condition to the surface source or aquifer. While a falling water table can be reversed to some extent by recharge, once subsidence has occurred, the long run damage to the aquifer may not be reversible (see Chapter 12). External costs should be included in the cost calculations for water and therefore in its price. To avoid excess utility revenue and profits from this policy would require a tax on the utility equal to its external cost. The higher water prices resulting from this policy might be politically acceptable if the tax were paid to a state environmental agency. A policy of full inclusion of costs in water prices is necessary for efficient use of resources.

3.5 Other Cost Concepts

Cost is often divided into fixed and variable components for operational analysis of the system. Pumps, wells, pipes, planning and engineering, and treatment systems are *fixed costs*. To the extent that they are on long-term contract, water department managers represent a fixed cost. Payments for energy, water purchase, chemicals for water treatment, and labor are typical *variable costs*. Variable costs are sometimes referred to as avoidable costs while fixed costs are frequently referred to as overhead costs or sunk costs. A real world test for fixed costs is to examine which costs would still exist if water production were shut down.

The economist and the accountant view long run cost differently. Accountants define the long run as a period longer than one year because accounting principles are driven by time constrained budgets and tax considerations. Economists define the long run as an indefinite period of time when all costs become variable. This is often referred to as the planning period.

Economists like to analyze *marginal cost*, the cost of the additional or last unit produced, which is also the change in total variable cost for a one-unit change in output.[6] Economic efficiency requires that marginal cost be equal to the value of the marginal unit produced of a good or service. In the short run, marginal cost is the change in variable cost associate with production of an additional unit (e.g. a thousand gallons of water or an additional connection) given the fixed system. Timmins (2002) points out that water is typically priced significantly below its marginal cost and this leads to inefficient exploitation of aquifers and other water sources.

If the water using community is growing, in the long run new sources of water, delivery systems, and treatment facilities must be acquired. The relevant marginal cost and marginal benefit calculation in this case is based on the entire cost, including both capital and variable costs, of the proposed new source of water. The relevant long run supply curve was shown earlier in Figure 3.5.

3.6 Economic Efficiency and Equity

In setting policy, economists are concerned about both *economic efficiency* and *equity*. Economic efficiency occurs when a resource is in its most productive use and the marginal value of satisfaction (marginal utility) produced by the resource is just equal to its marginal cost of production. A resource's opportunity cost is the value it could produce in its next best alternative use. Thus, a resource being used efficiently must cover its opportunity cost or be shifted to a higher and better use where its application will provide greater value. In a competitive market, economic efficiency occurs at the price and quantity where the supply curve intersects the demand curve. Conceptually, this requires the participation of many buyers and sellers in the market, as well as the absence of externalities. When these conditions are satisfied, the consumer of the

[6] Marginal cost is also the change in total cost as the change in total variable cost is the only change occurring as more units are produced in the short run.

marginal unit of water is paying its marginal cost of production. Furthermore, the value of the marginal unit of water to the consumer is just equal to its marginal cost.

Urban water markets do not meet these criteria because external costs are generally ignored and there are no competing sellers. When one purchases a new table, the marginal cost to the producer and the marginal benefit to the consumer are well known to the participants in the market[7]. In water, the high positive externalities of public health and the negative externalities from over drafting an aquifer, overdrawing water from a reservoir or stream, and disposal of wastewater make the marginal costs and benefits more obscure to those participating in the market.

Equity refers to the distribution of goods across and among people with various income levels. Equity is strongly influenced by taxation, regulation and public good pricing policies. Horizontal equity is achieved by equal treatment of individuals of the same income class. Vertical equity refers to the treatment of individuals across income classes. The achievement of vertical equity in urban water provision requires that the price schedule be in line with ability to pay and that the cost impact of regulations on low-income customers at least not be greater than that on high-income customers. The water provider must balance concerns for equity and efficiency to satisfy regulatory agencies, elected officials, and water users. Any change in rate structures or alignment of customer classes results in a redistribution of income. Thus, equity must be considered when rates, rate structures, and customer classes are changed.

3.7 Water Rates and Inflation

Proper pricing requires periodic rate increases to adjust for inflation. A simple way to achieve this result is to index water rates to the Consumer Price Index (CPI). The implementation of an annual automatic CPI adjustment would likely require considerable time, political risk, and effort on the part of those responsible, but once in place would no longer be an issue, since people are accustomed to similar adjustments in other programs such as Social Security and income taxes.

The initial drop in water purchases resulting from a one-time price increase during a period of inflation will gradually disappear as the *real price of water* declines. This often leads to the erroneous conclusion that price effects are only temporary. When comparing prices in different time periods the real price should be used. Real price is simply the price in each time period adjusted for inflation to some base year level. For example, during a period of relatively mild inflation of 4% per year, a 20% rate increase would be more than offset by inflation in less than 5 years, so within 5 years water use is likely to return to its prior level. It is not uncommon for water providers to set a price and then not adjust it for inflation over a long time period. This practice frequently results in the current real price falling below the previous real price. Table 3.2 shows the result in several Arizona cities of the failure to update water rates to keep up with inflation.

---

[7] Admittedly, in the case of the table there may be some small unknown internal and external costs associated with disposal of the table after its useful life. We choose to ignore these costs in making our argument.

Table 3.2 The Effect of Inflation on Typical Residential Water Bills for Selected Arizona Cities

Provider	Water Bills for 7,500 Gallons (28.39 m^3) per Month			
	Year Rate Established	Actual Water Bill	Adjusted for Inflation, 1st Quarter 1997	
			Adjusted Bill	Percent Reduction
Tempe	1992	$14.19	$12.31	13.2%
Chandler	1994	$18.06	$16.60	8.1%
Prescott	1991	$20.30	$17.18	15.4%
Casa Grande	1994	$25.95	$23.83	8.2%

Notes: Bills include fixed charges. Rates were set at different times during the year. Deflated by Consumer Price Index (base year 1983). Source: Black and Veatch (1997, 1995, 1993)

A local or regional consumer price index is the ideal deflation index to use when studying local water demand. Water price indexes should not be used for this purpose, since they reflect the cost of providing service rather than inflation in the general economy faced by the consumer. Local consumer price indexes are often kept by business research divisions of major universities, large city or regional governments, state governments (e.g.: Texas and California calculate price indexes for various regions of their state), and in the U.S., the Department of Labor, which calculates indexes for many major metropolitan areas. In the event that local data are not available, we recommend that a national consumer price index or GDP deflator be used. National deflators are usually available from the government or the central banking authorities. The formula for deflation of an economic series is:

$$\text{Real Value (inflation adjusted)} = \frac{\text{Nominal Value} \times 100}{\text{index number}} \qquad (3.1)$$

For example, assume the price of water was set at $2.40/1000 gallons in 1998 and remained unchanged through 2002. Also, assume the price index for 2002 is 134 based on 1998 prices, with the index in 1998 = 100. Therefore, the real or inflation adjusted price of water in 2002 dropped by $0.61 in real terms as shown in equation 3.2.

$$\text{Real Price in 2002 (1988 price level)} = \frac{\$2.40 \times 100}{134} = \$1.79 \qquad (3.2)$$

3.8 Conservation Policy--Regulation, Persuasion, Price Rationing or All Three?

Water providers seeking to reduce demand could seek government help in implementing laws governing landscape choices, forcing new homes to have low flow toilets, showerheads, and faucets, and limiting industrial or golf course use. Laws could also mandate use of reclaimed water for specific purposes when it is available. Such a strict regulatory environment would set arbitrary limits on water customers' activities that may not match well with their wants and needs. It would require a detailed set of regulations and high enforcement costs. Water use can also be limited by shutting off supplies for selected time periods. Such a policy has severe impacts on the quality of the water supply as well as on the activities of individual customers.

Price rationing, which is consciously using price to regulate demand, is both an alternative and a complement to regulation. An argument for price rationing over regulation is that customers are free to choose the quantities of water that they desire and can afford and that the customers are in a better position to judge their wants and needs than the authorities. It also avoids arbitrary decisions by regulators, and has no enforcement cost, since the increased water bill would be the enforcer of the desired reduction in water use.

The water provider may use persuasion to reduce the demand for water. Advertising can be used to convince customers that doing the right thing includes watering lawns during off peak hours, installing low flow devices, and having less water intensive landscaping. Some utilities offer subsidized low flow showerheads and technical advice on construction and landscaping. Costs are incurred by the water provider for advertising and supplies for these programs. However, if the marginal value of the water saved, which is its marginal benefit, is equal to or greater than the marginal cost of the program, the program may be worthwhile.

Economic efficiency dictates that the price of water be equal to its marginal cost and that the customer pays the marginal cost of consuming the last unit. However, municipal water is normally sold under monopoly conditions in which the marginal cost of the last unit of water sold is greater than the price. If the commodity charge is set equal to marginal cost, and both are below average total cost, losses are incurred that must be subsidized by local taxpayers. A private provider would not produce water services in this long run situation without a subsidy. On the other hand, if marginal cost is set equal to price and both are above average total cost a profit will be made. Many municipal water suppliers are prohibited by law from making a profit, a condition that in many cases leads to economic inefficiency. Most water providers do not attempt to use marginal cost pricing. Instead, they carefully calculate their cost of production, allocate it among various classes of customers, and then set prices high enough to cover total cost. The result is that price per unit is set equal to average cost of production. This is often referred to as cost of service pricing. Such a price satisfies the nonprofit condition, the fair return required by regulators for private providers and insures that losses will not be incurred. However, in most cases, this policy results in a price below marginal cost of providing water which encourages over use of water and is economically inefficient.

3.9 The Need for Metering

If water is not scarce, household and business units may be charged a periodic flat fee for service, which may increase with the size of the service connection, with no restriction on amount of consumption. The flat fee generally covers the cost of delivery and extraction of water. Most water utilities also impose a connection fee, which is a one-time charge to cover the water purveyor's cost associated with a new connection to the water system. This fee is generally levied in both metered and unmetered systems.

As water becomes scarce, metering of individual customers and imposing a commodity charge becomes necessary for economic efficiency and utility finances. Individual metering allows water use to be rationed by price. When water prices are sufficiently high, this encourages the use of substitutes for water, more water efficient technology for water consumption activities, and a conscious evaluation of water consumption habits. Conventional wisdom says that apartment renters with common meters consider water as a free good and, therefore, apartments with common meters are not subject to the effects of price rationing. However, even in this case, price rationing can be successful. With sufficiently high water prices, apartment owners will be induced to install less water using landscapes, improve swimming pool maintenance procedures, install low flow faucets and toilets, conduct leak check programs, and reduce the use of decorative fountains. Only the personal uses remain as a free good to the renters. Applying price rationing to these uses can be achieved by individual metering of apartments, which may or may not be cost effective.

3.10 Rate Structures

Most urban water providers impose a periodic meter charge combined with a uniform commodity charge, increasing block rates (often referred to as conservation rates), or decreasing block rates. The simplest rate structure is a two-part tariff with a fixed periodic meter charge, to cover non-water costs (e.g. meter reading, maintenance of infrastructure, administrative costs) and a uniform rate for each unit of water used, as shown in Figure 3.6. The fixed monthly charge can also be used to cover the difference between revenue raised by the commodity charge and total cost.[8] Utility managers often prefer a large monthly service fee since it reduces revenue fluctuations due to changing water sales. When the commodity charge is set equal to long run marginal cost, water production and use are economically efficient. However, the commodity charge is usually set equal to average cost to avoid profits or losses.

An objection to a uniform commodity charge for all household water is that it may be vertically inequitable. The question arises "should poor customers for a publicly provided necessity good like water pay the same commodity price as rich customers?" A uniform rate is inherently similar to a proportional sales tax in that its relative burden falls more heavily on low income people. Because higher income customers tend to purchase more water per capita and have more water using appliances than lower income customers, one could argue that both groups paying the same rate is inequitable to the lower income customers as the marginal uses of the additional water consumed

[8] The economics of the fixed charge and the uniform marginal price are different. Changing the fixed charge results only in an income effect which means the purchases of water and all other goods are affected. Raising the uniform price results in both an income and price effect. That is, water use falls both because water bills are higher, thus reducing spending on all other goods, and because the relative price of water increased.

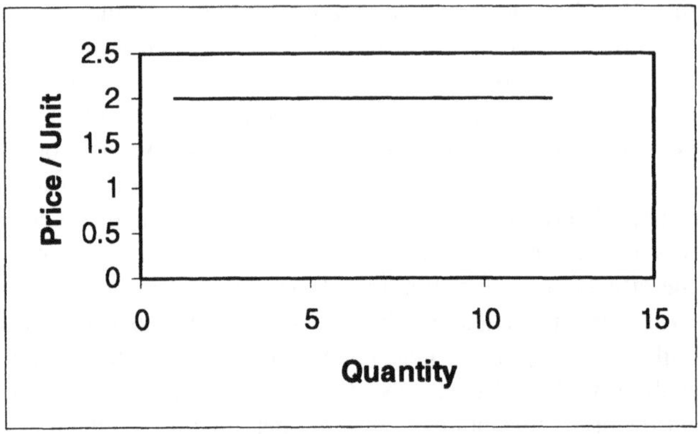

Figure 3.6 Uniform Commodity Charge

by the higher income groups are usually of lower value to society than those of the lower income group. For example, filling of a swimming pool or a decorative fountain by a rich person is generally considered of lower value to society than a poor person's use of water for cooking or hygiene purposes. If one believes that the marginal (incremental) value of money decreases as one acquires more money, one would argue that equity considerations support charging those with lower incomes a lower price and those with higher incomes a higher price.

3.10.1 INCREASING BLOCK RATES

The increasing block rate structure shown in Figure 3.7 is designed to remove some of the inequities described above. The first and lowest block is sometimes called a lifeline rate as it is intended to be equitable for low-income families who generally consume the lowest amounts of water. Each successive block beyond the first block has a higher marginal price and there is no particular limit to the number of blocks[9].

Changing from a single commodity price to a carefully designed increasing block rate structure can improve equity while at the same time being revenue neutral. Skill in rate design is needed to ensure that the spread of customers over the blocks will lead to the utility covering its costs. While potentially improving equity, increasing block rates introduce an element of economic inefficiency, since no more than one block can have the property of price equal to marginal cost. However, most real world rate schedules do not have any blocks with price equal to marginal cost. Besides this presumed lack of economic efficiency, another objection to increasing block rates is that consumers find them confusing and do not give expected responses to changes in the marginal block prices. A tariff with only two or three blocks is more likely to be effective than one with

[9] The ultimate extension of this idea is a continuously rising price as quantity increases. Given today's computing and monitoring equipment, this could become a possibility.

many blocks and may be more politically acceptable. Very high marginal rates for excessive consumption as a temporary measure during droughts have proved to be effective in reducing water usage.

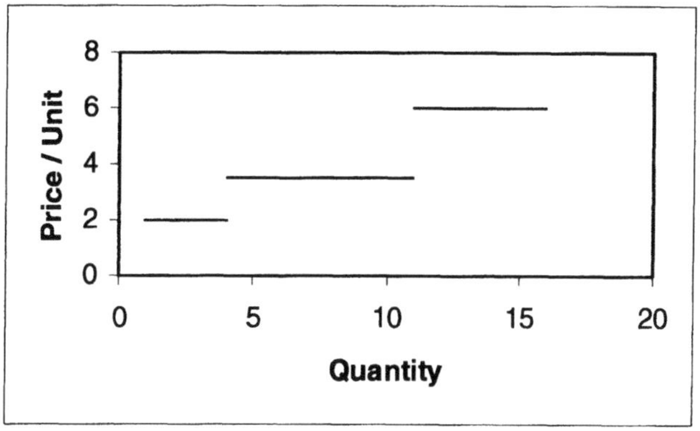

Figure 3.7 Increasing Block Rates

An additional feature of increasing block rates is that the lower priced lower use blocks offer a subsidy to consumption in the higher blocks, thus softening the effect of price on the consumers in the higher blocks. Thus, higher income people may support this subsidy to lower income people as they also benefit. However, the city of Tucson was forced to hold a recall election when prices were greatly increased and the blocks were not correctly designed to satisfy high-income high water using customers (Martin et. al., 1985). Furthermore, two studies of Tucson Water customers completed ten years apart revealed that most customers who believed that increasing block rate structures saved water also disliked the rate structure (Agthe and Billings, 1997). An alternative to the lifeline block is a government subsidy to low income people who apply and qualify for water bill payment assistance. However, given the generally low percent of income spent on water by most customers in high income countries, such a specific policy is of dubious value. While it may be appropriate to some systems in low income countries, the administrative cost is significant compared to simply adjusting the rate structure.

3.10.2 DECREASING BLOCK RATES

A cousin of the increasing block rate structure is the decreasing block rate structure where each successive rate block is lower in price than the previous block. This historically common rate structure, shown in Figure 3.8, is appropriate for water systems where water scarcity is not a problem. Also known as promotional rates, the rationale behind the structure is that large users allow the system to achieve economies of scale in water production and delivery. The reduced commodity charge for the large volume customers passes the economies of scale to those who helped create them. The lower price also protects the utility investment in infrastructure as these customers are frequently large enough to sink their own wells, develop their own surface water sources, or otherwise unhook from the system, leaving it with a high uncovered fixed obligation. The political danger to elected officials, utility rate setters, or regulators in this case is the small users who pay the highest marginal price objecting to the subsidy of the largest users.

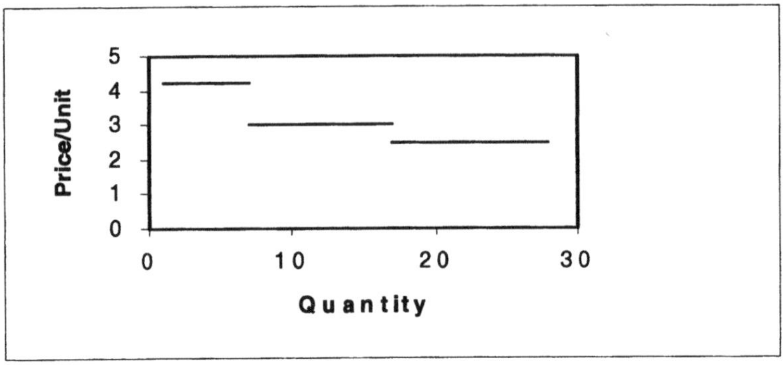

Figure 3.8 Decreasing Block Rates

3.10.3 CONSERVATION RATES

A marginal price of water that exceeds the utility's marginal cost of water provision is termed a conservation rate as its purpose is to conserve or reduce the amount of water consumed. The rational for conservation rates is that the social cost of environmental damage is not covered in the utility's cost structure and the marginal cost of the water, as measured by its replacement cost from the next most expensive source, is not accounted for by the utility. If the discounted values of these costs were added to the utility's production cost it is likely that greater economic efficiency in water use would be achieved. These rates could be directed at summer peaks or large users. The higher blocks in an increasing block rate structure could be conservation rates, as could a single rate set above marginal cost.

In colder climates, urban winter water consumption is mostly indoor use. However, water use increases greatly in the summer because of outdoor use for irrigation and swimming pools,[10] creating peaks in water use which can be a problem for the utility. The additional capacity required to provide for summer peak loads can be financed by imposing a summer surcharge set to cover the additional capital and operating costs of the peak load. A higher summer rate will also result in an overall reduction in peak summer usage. The utility could also impose a shoulder rate for the interim period between summer and winter. An alternative summer peak load rate schedule might impose a surcharge on each customer for use above that individual customer's average winter water use. For example, a surcharge might be imposed on all use exceeding 150 percent of the average water use for each customer in November,

[10]In warmer areas like California, Florida and Arizona which experience year around irrigation and swimming pool use, winter use includes some outdoor water use.

December, January, and February (northern hemisphere). Thus, each customer would have a customized base amount that when exceeded in summer will place him or her in a higher price block representing a customized conservation rate. However, employment of such a customized rate structure requires incentives against customers overusing water in the winter to achieve lower rates in the summer. Large volume summer customers like golf courses and city recreation areas would be attracted to this strategy. A variety of regulations to prevent this strategy have been implemented by water utilities, including direct negotiation with individual customers and water use standards based on water requirements for turf. The base quantity for a new customer or one that consumes very little or no water in the base period could be set equal to the average of all customers in that category.

3.10.4 CUSTOMIZED RATES BASED ON WATER RIGHTS

Another customized rate structure with some similarities to the customized summer rate structure has been proposed for the Monterey Bay area of California (Haddad, 2000). The base amount would be determined by the number of water rights held by each individual customer. These rights could be traded among owners as their needs for water change. A penalty rate would apply when any water customer's use exceeded their water rights holdings[11]. The rights would initially be assigned in some socially acceptable manner to get the market started. These assignments would probably reflect the preferences of those doing the assigning and would be highly political with winners and losers. However, after its start the water rights market could operate as a free market, discussed at the beginning of this chapter. While interesting, the practicality of such a program is unknown. Of particular concern is the likely impact on economic development, as developers may have to acquire water rights in the market, depending on how many rights are initially assigned to undeveloped land. If the trading volume is too small or transaction costs too high, the market would be less efficient than envisioned by its planners. Big companies with high paying jobs and big developers might not want to signal their plans by the purchase of water rights and might locate elsewhere, causing a lower rate of growth in the local economy.

3.10.5 SEWERAGE USAGE FEES

In many urban areas water and sewer services are provided by the same utility and a fee for sewer use is added to periodic water bills. This fee could be a set amount per connection, an amount based on water use, or determined by measured sewage flows. However, with currently available technology, it is more efficient to base sewage charges on water use for most customers. A typical sewage use fee is based on winter water demand--(i.e.: November through February in northern climates)--as outdoor uses are few in that season and most of the water is assumed to flow into the sewer system.

[11] While one penalty rate is suggested, it is an easy extension to see that any number of successive increasing penalty rates (to impose higher charges on more flagrant over use) could be imposed instead of a single penalty rate.

The sewage use fee is then added to water bills for each billing period that follows the winter base period until a new winter base period is obtained. Sewage use fees based either on winter water use or direct measurement are likely to encourage more efficient use of water resources, since the fee will vary with the quantities used. In contrast, flat fees provide no incentive for efficiency.

The cost of disposing of sewage is part of the cost to society of using water, so combined water-sewer bills are generally an efficient means of matching customer use rates with costs of provision. However, in many areas sewage collection and treatment costs are financed through general taxation and therefore there is no incentive for efficient resource use. In very low income housing areas, however, general tax financing may be the only way in which this essential public health service can be provided.

3.11 Reclaimed Water Use

Reclaimed water is an effective substitute for potable water in many uses, but its potential contribution to solving the water supply problem is often unrealized. Reclaimed water use requires an expensive dual water delivery system, additional treatment of sewage, and effective marketing to overcome its negative image. Charging the marginal or average cost of producing reclaimed water when it is introduced may limit the number of customers and prevent achieving production levels necessary for internal economics of scale. A low promotional price may be necessary to attract users in order to obtain the benefits from substitution of reclaimed water for potable water. These benefits may include not having to develop new expensive sources of potable water, reduction in environmental damage caused by overdrafting an aquifer and reduced adverse environmental impacts of typical sewage disposal practices. It may be better for the water provider to subsidize initial customer use so that production volume can later reach economically efficient levels. The subsidy could be funded by general tax revenue or by raising prices for potable water.

3.12 Conclusion

Most of this chapter was devoted to defining economic concepts and showing how they can be used in urban water supply management. However, once the concepts of demand, cost, supply, economic efficiency, and equity were established, the use of price rationing through conservation rates was discussed. Several rate structures and their implications for equity and efficiency were examined: flat rates, increasing block rates, decreasing block rates, and flat rates combined with a high penalty rate for "excessive" use. The flat rates are likely to be the most economically efficient but not vertically equitable. Vertical equity can be achieved with block rates but, at most, only one block will have an economically efficient price.

The penalty rate can be very effective in managing drought conditions, but it is unlikely to be politically acceptable under normal conditions. To achieve a desired level of conservation it is likely that conservation rates will be combined with other measures such as mandates for low water use fixtures in new or even re-sold buildings and incentives for converting to water conserving landscaping. Finally, if the conservation program cannot be "sold" politically to the customers, it will "not fly."

References

Agthe, D.E. and R.B. Billings, 1997. "Equity and Conservation Pricing Policy for A Government Run Water Utility," *Journal of Water Supply Research Technology-Aqua*, 46 (5):252-260.

Black and Veatch, 1997. *Arizona Water/Wastewater Survey 1997*, Phoenix, Arizona.

Black and Veatch, 1995. *Arizona Water/Wastewater Survey, 1995*, Phoenix, Arizona.

Black and Veatch, 1993. *Arizona Water/Wastewater Survey, 1993*, Phoenix, Arizona.

Haddad, B.M., 2000. "Economic Incentives for Water Conservation on the Monterey Peninsula: the Market Proposal," *Journal of the American Water Resources Association*, 36(1):1-13.

Johnson, D., 1988. "Down the Drain," *The Los Angeles Times*, Feb. 2, Sect 5:1.

Martin, W.M., N.H. Laney, and A.W. Griffin, 1983. *Saving Water in a Desert City*, Resources for the Future, Inc., Baltimore.

Timmins, C. 2002."Measuring the Dynamic Efficiency of Regulator's Performances: Municipal Water Utilities in the Arid West," *Econometrica*, 70(2):603-629.

Suggestions for Additional Reading

Agthe, D.E., and R.B. Billings, 1987. "Equity, Price Elasticity and Household Income Under Increasing Block Rates for Water," *American Journal of Economics and Sociology*, 46 (3):201-212.

Billings, R.B., and D.E. Agthe, 1980. "Price Elasticities for Water: A Case of Increasing Block Rates," *Land Economics*, 56(1):73-84.

Billings, R.B. and C.V. Jones, 1996. Chapter 9, *Forecasting Urban Water Demand*, American Water Works Association, Denver, Colorado.

Foster, H., and B. Beattie, 1979. "Urban Residential Demand for Water in the United States," *Land Economics*, 55(1):43-58.

Haddad, B.M., *Rivers of Gold: Designing Markets to Allocate Water in California*, Island Press, Washington, D.C.

Howe, C.W., and F.P. Linaweaver, 1967. "The Impact of Price on Residential Water Demand and Its Relation to System Design and Price Structure," *Water Resources Research*, 3(1):13-32.

Mansfield, E., 1994, *Applied Microeconomics* 2nd. Edition, W.W. Norton and Company, New York, New York.

Martin, W.E., and J.F. Thomas, 1986. "Policy Relevance in Studies of Urban Residential Water Demand," *Water Resources Research*, 22(13):1735-1741.

Neufeld, J.L., and J.M. Watts, 1981. "Inverted Block or Lifeline Rates and Microefficiency in the Consumption of Electricity," *Energy Economics*, 3 (2):113-121.

Stigler, G., 1966. *The Theory of Price*, The MacMillan Company, New York.

Chapter 4

Water Supply Economics

Benedykt Dziegielewski
Southern Illinois University

4.1 Introduction

Urban water supply and sanitation systems play an important role in maintaining public health and safety. Many public water utilities, which build and operate water systems, also provide wastewater collection and treatment service. This chapter is focused primarily on the costs of water supply and, to a lesser extent, on the economics of wastewater treatment and reclamation.

The discussion begins with a brief background on the water supply and sanitation industry in the United States with the help of statistical data developed by periodic surveys of public water supplies and sanitation districts that are conducted by the U.S. Environmental Protection Agency. The costs of water supply are then introduced from the perspective of full cost to society and full value in use. The EPA data also are used to characterize the structure of water supply costs in terms of average expenses per unit volume and various cost components.

Published data on the investment, operation and maintenance costs for five different water supply alternatives are reviewed. These include conventional groundwater and surface water supplies and three unconventional options including water demand management, water reclamation and desalination.

The last section deals with the economic concepts of costs and their implication for water pricing. A case study of Phoenix, Arizona, is cited to illustrate the cost of capacity expansion and implied marginal cost-based prices of water.

D.E. Agthe et al. (eds.), Managing Urban Water Supply, 45–70.

4.2 Water Supply and Sanitation Systems

Modern public water-supply systems are designed to provide "on-demand" delivery of highly purified water to homes and businesses. Nearly all water delivery systems are also designed to accommodate seasonal and daily variability in water use and satisfy high-flow demands for fire protection. While the major part of water delivered is for non-potable uses, all water is purified to meet the drinking water standards of the Safe Drinking Water Act (SDWA) Amendments of 1996. The need to meet this high water quality standard is a major reason behind the increasing cost of urban water supply. In the United States, the water supply industry consists of approximately 54,000 community water systems and 214,000 non-community systems (EPA, 1997). According to the 1999 survey of community water systems, the total national need for drinking water investments is $150.9 billion for the next 20 years, including $102.5 billion needed immediately to ensure the continued provision of safe drinking water (EPA, 2001).

The provision of water is a very capital-intensive industry, second only to the energy industry. Total system assets range from $31,900 in systems serving less than 100 persons to $63.1 million in systems serving between 50,000 and 100,000 persons. Capital investments include outlays on water infrastructure, which include water sources (wells or reservoirs), transmission lines, treatment facilities, water distribution lines and storage. The value of total physical assets of all 54,000 community water systems is substantial. Based on the 1997 EPA survey data, the estimated total assets are $221 billion; a significant percentage of the total value of infrastructure in all U.S. utilities that in 1999 was reported to be $1,311 billion (BEA, 2000). In terms of investments per revenue dollar, the water industry is probably the most capital intensive of all utilities. Table 4.1 shows that the ratio of assets to total annual revenue ranges from 2.17 to 5.83 per dollar of revenue, depending on system size.

Table 4.1 **Total Assets in Public Water Supply Systems (Median Values)**

System Size (people served)	Number of Systems	Total Assets[a] $1000	Ratio of Assets to Revenues
100 or less	31,688	15.2	2.17
101-500		100.5	3.33
501-1,000	14,149	254.5	4.19
1,001-3,300		860.7	4.37
3,301-10,000	4,458	3,069.4	5.14
10,001-50,000	3,416	10,701.0	5.22
50,001-100,000		39,868.9	5.64
Over 100,000	353	112,565.5	5.83

[a]Total assets include everything the utility owns such as cash, accounts receivable, inventories, property, plant and equipment. Source: EPA, 1997.

The majority of water delivered to homes and businesses is returned in the form of municipal wastewater. According to the U.S. EPA, in 1996 there were about 20,000 publicly owned municipal wastewater collection systems and approximately 16,000 wastewater treatment facilities which served approximately 190 million people (EPA, 1997). An additional 3,000 facilities are to be built in the near future.

The total 1998 value of fixed assets consisting of sewer system structures was estimated at $308.4 billion (Herman, 2000). The total national investment needs for wastewater systems by 2016 are estimated to be $139.5 billion which includes $44 billion for wastewater treatment, $10.3 billion for upgrading existing collection systems, $21.6 billion for new sewer construction, and $44.7 billion for controlling combined sewer overflows (EPA, 1997).

The high capital investment needs are the major determinant of the cost of water supply. Costs are even higher when viewed from the societal accounting perspective, which in addition to the capital and operations and maintenance (O&M) costs also considers opportunity cost and other economic and environmental externalities. Generally, the cost paid by consumers is significantly below the full societal cost of water supply provision and wastewater disposal.

4.3 The "Full Cost" of Water

Rogers, Bhatia and Huber (1998) distinguished three levels of cost that are associated with the human use of water: full supply cost, full economic cost, and full cost of water consumption (or use). The composition of each level of cost is illustrated in Figure 4.1.

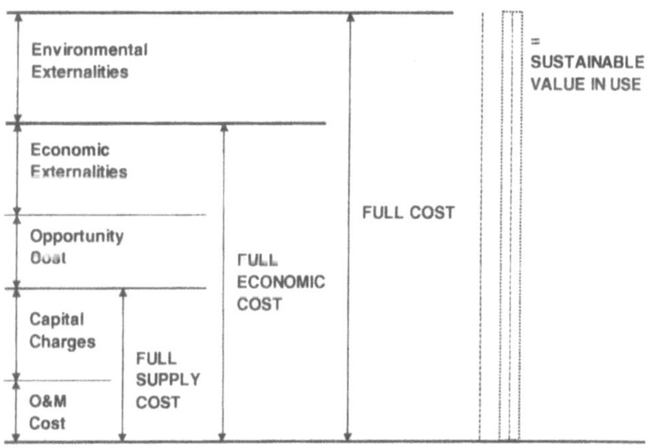

Figure 4.1 **Framework for Evaluating the Full Cost of Water (Source: Adopted from Rogers, et al., 1998)**

4.3.1 "FULL SUPPLY COST" OF WATER

The "full supply cost of water" includes all expenditures that are necessary to supply water to the consumer. Because extensive infrastructure is required to supply water to individual users, the cost of water supply is composed of capital and O&M costs. Capital costs include the initial and periodic expenditures of constructing water intakes, transmission lines, treatment plans, storage reservoirs and distribution lines. Because the capital investments in water supply are infrequent and involve financing, a convenient measure of these costs is the sum of depreciation (i.e., capital consumption) and interest paid on capital improvement loans. Rogers et al. (1998) note that there is some disagreement about the proper calculation of capital charges. Traditional methods use a backward accounting approach and include only the costs associated with repaying the historical investments. Newer approaches use forward-looking accounting and consider the cost of replacement of the physical assets and the potentially increasing cost of new additions to the capacity of supply sources.

The O&M costs are expenditures incurred in day-to-day operations of a water supply system. They represent the cost of various inputs, which are required to produce and deliver water including administrative cost. Typical inputs include electricity, chemicals, labor, and materials and equipment for system repairs and maintenance. These costs can easily be measured using the transactions recorded in the utility's accounting books.

4.3.2 FULL ECONOMIC COST

The "full economic cost" of water use by a specific sector (e.g., municipal and industrial water supply), includes the O&M cost and the capital charges as well as two other categories of economic costs that are external to the utility that supplies water. These are the opportunity cost and economic externalities.

The opportunity cost arises when the use of water by one user or sector deprives another (potential) user of putting the same water to a higher value use. The opportunity cost is zero when there is no shortage of water and all uses (including in-stream and off-stream uses) are fulfilled. The opportunity cost is related to the allocation of water resources among alternative uses. A non-zero opportunity cost indicates that water is misallocated among economic sectors or individual users.[12]

Economic externalities may be either negative or positive. These arise when some costs associated with the use of water are borne by another entity (a negative externality) or another entity receives some economic benefit (a positive externality). For example, intensive pumping of groundwater by a municipality may significantly lower the groundwater table, thus increasing the cost of pumping water from private wells or creating the need to drill deeper wells. Conversely, a utility that imports water for artificial recharge of an aquifer may help reduce pumping costs from private wells. Significant negative externalities are often associated with the release of polluted

[12] Ed. note: A different definition of opportunity cost appears in Chapter 3. Opportunity cost in Chapter 3 refers to the total cost of use of a resource while in the current chapter it refers to the incremental cost between two possible uses.

effluents by upstream users. Economic externalities are most often discussed in the context of over-extraction or pollution of water supply that increases the production or consumption costs of others.

4.3.3 FULL SOCIETAL AND ENVIRONMENTAL COST

Rogers et al. (1998) make a distinction between economic and environmental externalities because the latter are more difficult to assess than the former. They define environmental externalities as those associated with public health and ecosystem maintenance. As shown in Figure 4.1, the full cost of water use includes not only O&M and capital costs of water supply, but also opportunity cost and economic and environmental externalities. The full cost of water is important when considering the criteria for the sustainable use of water resources. Ideally, the full cost of water should equal the sustainable value in use, which is the sum of the economic and intrinsic values shown in Figure 4.2. According to Rodgers et al. (1998), the economic value of water includes the value derived by the users of water (as determined by the willingness to pay or marginal value of product), net benefits from return flows (e.g., groundwater recharge through irrigation), net benefits from indirect use (e.g., water for flora and fauna in diversion canals), and adjustment for societal objectives (e.g., poverty alleviation or employment). In addition to these economic values, the full value of water includes its intrinsic values such as stewardship, bequest value and existence value. For economic equilibrium, the value of water should just equal the full cost of water for its marginal use, so that the total full value is higher than the total estimated full cost since non marginal uses will have values in use exceeding full cost.

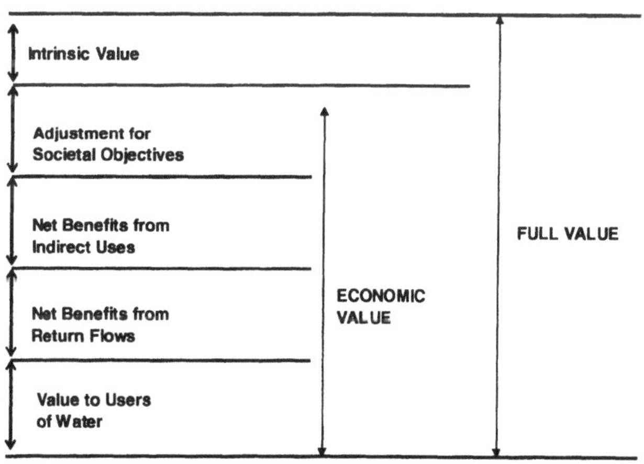

Figure 4.2 Components in the Value in Use of Water
(Source: Adopted from Rogers et al., 1998)

The "full cost pricing" concept implies that the charges levied on the users of water should recover the full cost of water consumption. In the real world, the charges for water are significantly below the full cost. Table 4.2 shows examples of the full cost, value in use and water tariffs for three user sectors in the Subernarekha River Basin in India. The data in Table 4.2 clearly indicate that water users in the Subernarekha study area pay only a small fraction of the full cost of water supply. Urban water users pay only 18 percent of the O&M and capital cost (estimated at 6.6 cents/m^3) and only 2.6 percent of the full cost of water consumption. If "full cost" pricing were to be implemented, the cost of water to consumers would increase dramatically and the quantities of water that are presently consumed would significantly decrease.

Table 4.2 Example of Value in Use, Costs and Water Tariffs Subernavekha River Basin, India (in cents per m^3)

Description	Irrigation	Urban Use	Industrial Use
Full value in use	9.7	25.0	260.0
Full cost[a]	65.0	46.7	46.7
Water tariff	0.1	1.2	2.5
Ratios			
Value/Cost	0.149	0.535	5.56
Tariff/Cost	0.002	0.026	0.054
Tariff/Value	0.010	0.048	0.001

Full cost per cubic meter for irrigation includes: 0.2 cents in O&M cost, 3.8 cents in capital cost, 1.5 cents in pumping cost, and 59.5 cents in opportunity cost. Full cost of urban and industrial supply includes: 6.6 cents of full supply cost, 9.7 cents in opportunity cost, 1.4 cents in economic externalities, and 29 cents in environmental externalities.

Source: Rogers, Bhatia and Huber (1998)

While some economists propose new pricing strategies that would include the opportunity cost [e.g., a marginal opportunity cost pricing approach suggested by Warford (1997)], the first step would be to devise water rates that recover the actual utility expenses and provide a signal to consumers about the incremental cost of additional supplies, for example through the use of marginal cost pricing.

4.4 Structure of Water Supply Costs

4.4.1 AVERAGE EXPENSES PER UNIT VOLUME

The average (per unit) cost of water supply varies greatly among the 54,000 U.S. community systems. Major factors that contribute to this variability include system size, type of water source, and quality of raw water. Generally, larger systems can produce water at lower cost than smaller systems. As shown in Table 4.3, the average expenses (costs) per 1,000 gallons of water produced decline with increasing system size (with the exception of the 501-1000 size category). When the costs are recalculated separately for small and large systems, the median value of operating cost in small systems (i.e., those serving less than 3,300 persons) is $2.17 per 1000 gallons ($0.57/m^3) as compared to $1.65 per 1000 gallons ($0.44/m^3) in large systems. The declining average cost per 1000 gallons suggests that water utilities benefit from significant economies of scale. However, this relationship may not be straightforward because of the interaction between economies of scale in water treatment and diseconomies of dispersion in water transmission and delivery in small water systems (Boisvert and Schmidt, 1996).

Table 4.3 Total Annual Expenses and Revenues in U.S. Public Water Supply Systems (Median Values)

System Size (people served)	Number of Systems	Total Expenses $/capita	Total Revenues $/capita	Total Expenses/ 1000 gal.	Total Revenue/ 1000 gal.
100 or less	31,688	60	79	2.24	2.88
101-500		72	83	2.11	2.85
501-1,000	14,149	81	93	2.34	3.08
1,001-3,300		88	97	2.04	2.55
3,301-10,000	4,458	89	99	1.98	2.45
10,001-50,000	3,416	77	100	1.53	2.09
50,001-100,000		75	93	1.48	1.96
Over 100,000	353	76	90	1.31	1.73

Total expenses include all operating expenses, interest payments, depreciation, and other expenses (excluded are any payments of principal, capital improvements and advanced sinking-fund contributions). 1000 gallons = 3.785 m^3. Source: EPA, 1997.

The cost also depends on the type of water source. Water obtained from surface sources such as rivers or lakes often requires costly treatment to meet the SDWA standards. Because of the economies of scale in water treatment, surface water sources are used primarily by larger systems. Small systems tend to purchase water from larger systems with surface water supplies or to rely on their own wells.

Table 4.4 shows the median values of total expenses per 1000 gallons of water by system size and by primary water source. While unit cost declines with increasing system size for all three types of water sources, it is significantly lower among systems using groundwater than surface water. For all 1,539 systems in the EPA sample, the median expense is $2.18/1000 gallons ($0.58/m^3). For surface water systems, the median cost is $1.94/1000 gallons ($0.51/m^3), which is 11 percent higher than the median expense for groundwater systems. For very small systems (i.e., serving less than 500 persons) this difference approaches 50-60 percent. Systems that rely on purchased water show the highest expenses per unit of water.

4.4.2 COMPONENTS OF WATER SUPPLY COST

A major part of the cost of water supply is in the form of fixed cost. Debt service and debt-related expenses are approximately one third of total annual expenses. In addition, two thirds of routine operating expenses are fixed: that is, they do not depend on the quantity of water produced. Table 4.5 shows the structure of operating and non-operating expenses in a sample of water systems (EPA, 1997). Salaries and fringe benefits of system personnel represent the largest category of expenses. Four additional line items within operating expenses are also likely to be fixed. Only two categories, "energy, chemicals and supplies" and "water purchase expense" are expected to vary with the volume of water produced. For all systems, these variable costs represent about 33 percent of total operating expenses. The data in Table 4.5 also show that the structure of the annual expenses remains approximately the same regardless of system size.

Table 4.4 Total Expenses by Primary Water Source
(Median values in $/1000 gallons)

System Size (people served)	Groundwater Systems (n=756)	Surface Water Systems (n=376)	Purchased Water Systems (n=407)
100 or less	2.13	3.50	4.15
101-500	1.94	2.91	3.90
501-1,000	2.19	2.68	3.37
1001-3,300	1.72	2.32	3.60
3301-10,000	1.58	1.95	2.77
10000-50,000	1.40	1.32	2.23
50001-100,000	1.28	1.30	1.80
100,000 or more	1.13	1.29	1.64
All system sizes	1.75	1.94	3.22

(1000 gallons = 3.785 m^3) Source: EPA, 1997

4.5 Cost of Water Supply Alternatives

Water supply systems which need to expand their water supply capacity to serve new growth, or replace the current sources of supply due to declining water quality, usually consider a number of conventional and unconventional alternatives. The conventional sources include direct river withdrawals, on-stream reservoirs, pumped-storage reservoirs (also known as skimmer or side-stream), or various types of groundwater sources (i.e., direct withdrawal wells, infiltration galleries, or artificial storage and

Table 4.5 Percentage Distribution of Operating and Non-operating Expenses

Cost Category	Percent of Total Cost by System Size		
	3,300 or less	3,300 or more	All systems
Operating expenses[a]			
Direct compensation and benefits	28.7	33.8	31.0
Energy, chemicals and supplies	25.6	19.9	23.0
Outside labs and contractors	12.7	3.6	8.6
Depreciation expenses	8.3	11.9	9.9
Water purchase expense	8.4	11.6	9.8
Taxes and payments in lieu of taxes	3.0	3.7	3.4
Other operating expenses	13.2	15.5	14.2
Total	100.0	100.0	100.0
Non-operating expenses[b]			
Interest payments	87.6	90.9	89.4
Other debt service expenditures	6.6	5.6	6.0
Other expenses	5.8	3.4	4.5
Total	100.0	100.0	100.0

[a] Based on a sample of 1,526 systems [b] Based on a sample of 927 systems Source: EPA (1997)

recovery wells). Unconventional sources include long-term reductions in water demand (i.e., water conservation), water reuse (both direct and indirect with dual distribution systems), and desalination of brackish or sea water. The final selection of one or more alternatives will depend primarily on costs and on characteristics such as the available yield, water quality, and reliability of the supply source. The following sections consider the costs and practicality of several alternative water supplies.

4.5.1 SURFACE WATER SUPPLIES

Surface water from rivers and lakes is the primary source of supply for most large systems. In the U.S., nearly 5000 systems, or 10 percent of all systems, obtain all or almost all of their water from surface sources. These large systems serve more than half of the U.S. population. Although some systems rely on direct river withdrawals (i.e., run-of-the river supplies), the most common type of surface supply is one that utilizes storage in man-made reservoirs.

There are more than 75,000 dams and reservoirs in the U.S. with a total storage capacity of about 860 million acre-feet (Frederick, 1995). Many of these reservoirs are used to provide urban water supply. The cost of reservoir storage varies greatly among individual sites because of topographic differences and the economies of scale. Table 4.6 compares average costs of water storage for municipal and industrial water supply in 63 reservoirs built and operated by the U.S. Army Corps of Engineers (USACE., 1998). These costs range from $103 per acre-foot in the Southwestern Division to $945 in the North Atlantic Division. The current cost of new reservoirs is likely to be higher than those built in the past because most of the best and cheapest reservoir sites have already been developed. Frederick (1995) reports that for the 100 largest dams in the U.S. the average reservoir capacity produced per cubic yard of dam declined 35-fold between the 1920s and 1960s. Another important reason for the rising cost of surface water supplies is the high environmental cost of storing and diverting water.

Table 4.6 Investment Cost of Reservoir Storage Space

USACE Division	Number of Reservoirs	Total Storage (acre-feet)	Total Investment ($1,000)	Investment Per Unit Volume	
				$/acre-foot	Cents/ 1000gal.
North Atlantic	7	142,450	134,632.9	945.12	0.29
South Atlantic	10	217,366	117,569.9	540.88	0.17
Great Lakes and Ohio River	17	631,409	70,389.2	111.48	0.03
Mississippi Valley	6	369,650	41,660.6	112.70	0.03
Northwestern	12	807,240	111,655.4	138.32	0.04
Southwestern	63	6,885,616	714,150.9	103.72	0.03
South Pacific	2	470,900	104,914.9	222.80	0.07
All Divisions	117	9,524,631	1,294,974	135.96	0.04

Source: USACE, 1998

4.5.2 GROUNDWATER SOURCES

Groundwater is used as a primary source of supply by the 80 percent of community water systems that serve approximately one half of the U.S. population. Although groundwater is ubiquitous, its suitability for urban water supplies is determined by the maximum yield of production wells. Also, aquifers that underlie urban areas are susceptible to contamination and in some cases cannot be used because of poor water quality. However, because groundwater supplies are inexpensive to develop and maintain, they are the primary water source for small community water systems. Only six percent of small systems (i.e., serving less than 3,300 persons) rely on surface water.

The cost of a new well depends primarily on its depth and diameter. Engineering cost estimates for drilling and installing wells range from $12 to $18 per inch of diameter-foot of well depth. Higher costs apply to wells that are drilled in the rocky material. The cost of the wellhead including electric motor and pump is about $8,000 per inch of well diameter. At these costs, a small groundwater supply yielding 250 thousand gallons per day can be developed and operated at a cost of 20 cents per 1,000 gallons or 5.3 cents per m^3 (LCRWPG, 2000). However, the cost rapidly increases if the well water requires costly treatment or if wells with sufficiently large yield can only be found some distance from the community, thus adding the cost of a transmission pipeline and pumping.

4.5.3 WATER DEMAND REDUCTION MEASURES

Piped urban water supply systems are designed to deliver water under continuous pressure “on demand” as the system operators have no direct control over the quantity of water taken out by the customers. Consequently, water demand is usually taken as a given quantity that must be matched with supplies. However, past experience with water shortages caused by droughts indicates that water demands are not "given" but can be increased or decreased through various water management interventions. During a number of recent drought episodes, water providers were able to restrain water demands by persuading water users to temporarily reduce their water use while allowing distribution systems to remain fully pressurized. During the 1976-77 drought in California, restrictions on selected uses of water and water rationing plans were reported to reduce water use by up to 65 percent (Hoffman et al., 1979). This experience has led to the development of various programs for achieving long-term reductions in water use or loss through adoption of efficient water-using practices and technological innovations in the form of water-saving fixtures, appliances and other conservation techniques.

There are many methods of reducing water use or losses, and several possible vehicles through which these methods can be disseminated and implemented. Any activity, practice, technological device, law or policy that can potentially reduce water use may be considered a conservation measure. Examples of technologies and efficient water use practices that can be employed to reduce water use in urban areas are listed below (Dziegielewski, et al. 1993):

- Domestic uses: low-flow showerheads, shower-flow restrictors, toilet-tank inserts, faucet aerators, low- and ultra-low flush toilets, dual flush toilets, insulation of hot-water pipes, horizontal axis washing machines, low-pressure supply connections, pressure-reducing valves, water-efficient landscape designs, efficient landscape irrigation practices and other devices
- Industrial uses: counter-flow washing and rinse systems, reuse of process water, recirculation of cooling water, ozone treatment for cooling towers, treatment and reuse of blowdown, water recycling
- All uses: metering of water use, rehabilitation of water delivery systems, leak detection and repair, pressure reduction in distribution systems

These individual measures can serve as building blocks in the formulation of an "implementable" demand management program, which can be viewed as the counterpart of supply-side alternatives. Table 4.7 gives examples of typical demand management programs. Once the programs are designed, their long-term impacts on water demand and their cost-effectiveness can be evaluated. Each program will result in different water savings, as well as different magnitudes and incidence of costs. By the early 1980s, the attractiveness of long-term demand management measures had begun to be noticed by urban water supply agencies. For example, planners in one of the largest water providers in the United States have estimated that by 2010 water conserved through increase efficiency of water use in the urban area of southern California will constitute the largest single source of additional water (Dziegielewski, 1992).

Table 4.7 Demand Management Programs and Implementation Approaches

Program Category	Specific Implementation Program
Public Education	Primary and secondary school programs Promotional campaigns and events Mass media advertising campaigns Dissemination of information through personal contacts Outreach programs to educate water users and help them install conservation hardware Xeriscape[a] garden demonstration
Management Programs	Pricing and rate-making policies Meter testing and replacement programs Leak detection and repair program Distribution system audit program Tax incentives, subsidies and rebates for adoption of conservation measures Social conservation incentives and disincentives program
Regulation	Government enforced performance criteria and standards Local conservation codes and ordinances Water use restrictions and bans during emergencies

[a] Xeriscape refers to low water use landscaping

Data on costs of demand management alternatives are readily available. For example, a cost analysis performed for a water utility in Eugene, Oregon, showed unit costs ranging from U.S. \$0.01/1000 gallons (\$0.0026/m^3) to \$0.26/1000 gallons (\$0.069/m^3) with the higher end cost for commercial water audits (Dziegielewski and Davis, 1997). Similarly, Howe and White (1999) calculated levelized costs of water savings for 15 different measures for a demand-management program in Sydney, Australia, which would achieve demand reductions reaching 38 percent by the year 2011 as shown in Table 4.8. The levelized costs to the community ranged from AU \$0.0014/m^3 for implementing a showerhead performance standard to AU \$0.70/m^3 for a washing machine rebate program.

The results from Sydney indicate that significant reductions in average per capita water use can be achieved at very low unit cost of less than one cent (Australian) per cubic meter. However, the unit cost increases fairly rapidly for additional programs. Figure 4.3 shows the relationship between the accumulated savings and levelized cost based on data from Table 4.8.

Table 4.8 Water Savings and Costs of Demand Reduction Programs in Sydney, Australia

Demand Management Option	Demand Reduction (lpcd*)	Levelized Unit Cost (AU$/m^3)
Showerhead performance standard	8.6	0.0014
Price increase (AU$0.10/m^3 over two years)	1.9	0.0018
Clothes washer performance standard	3.5	0.041
Permanent low-level outdoor water use restrictions	1.8	0.063
Showerhead rebate (AU$10)	0.7	0.14
Residential indoor audit and retrofitting (AU$15 fee)	3.4	0.19
As above but free for low income households	1.5	0.25
Active leakage control	7.2	0.30
Industrial and commercial water use audits	2.9	0.42
Hotel audits	1.3	0.42
Outdoor water use promotion/advisory program	0.2	0.49
Industrial reuse project #1	2.3	0.53
Industrial reuse project #2	1.8	0.65
Outdoor irrigation system audits	0.3	0.67
Washing machine rebate (AU$150)	0.4	0.70
Total demand reduction/weighted average cost	37.8	0.22

* lpcd = liters per capita per day. 1 gallon= 3.785 liters. Source: Adapted from Howe and White (1999)

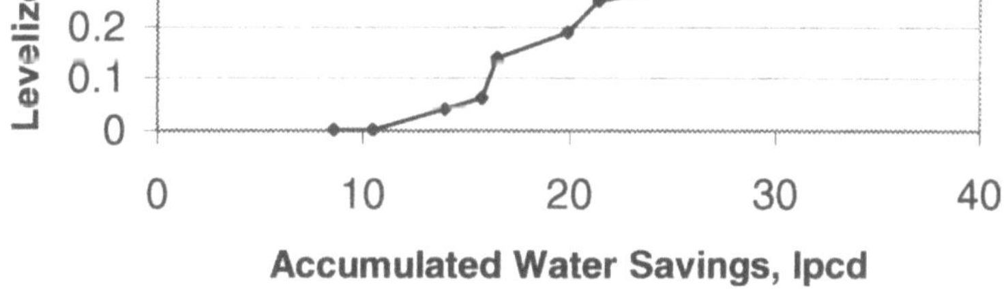

Figure 4.3 Accumulated Savings vs. Levelized Cost
Based on data from Howe and White, 1999

4.5.4 WATER REUSE AND DUAL DISTRIBUTION

Most of the urban end uses of water, especially landscape irrigation, industrial cooling, or toilet flushing, do not require water of drinking quality, so treated municipal wastewater can be used for such non-potable purposes. Furthermore, available water purification processes can convert municipal wastewater into reclaimed or repurified water which exceeds drinking water quality standards. However, public resistance to direct reuse of reclaimed water (i.e., introducing it directly into the water distribution system) limits the utilization of such water to non-potable uses.

Asano (2001, p. 3) clarified several terms that pertain to water reuse such as wastewater reclamation, direct and indirect use and water recycling. He noted "wastewater reclamation is the treatment or processing of wastewater to make it reusable, and water reuse is the use of treated wastewater for beneficial purposes such as agricultural irrigation or industrial cooling." Direct reuse does not necessarily mean potable use; it simply implies that reclaimed water is delivered by pipes to intended applications. Indirect reuse means discharge of reclaimed water to receiving waters for assimilation and withdrawal downstream. Finally, water recycling denotes a situation which involves only one use or user whose effluent is captured, treated and redirected back into the same system. Indirect reuse and water recycling have been practiced for a long time. Asano (2001) has compiled a list of 19 water reclamation and reuse projects dating back to the 1912 application of treated wastewater for watering lawns and supplying ornamental lakes in the Golden Gate Park in San Francisco, California. Florida and Arizona are the leading consumers of reclaimed water in the U.S. In 1995, their combined use was 2.84 million m^3 or 750 million gallons per day (mgd), which represented 74 percent of the total U.S. use of reclaimed water (Solley, et al. 1998).

The economics of water reuse and on-site recycling are driven by environmental and health regulations. Environmental regulations require that wastewater is treated before it is released to the environment. In many cases, three stages of treatment are needed (i.e., primary, secondary and tertiary) and the cost of such treatment can reach $1.00 per 1000 gallons ($0.26/m^3). The marginal cost of the (dual) conveyance and distribution system for bringing reclaimed water to users is the principal cost of most water reuse projects. Based on California experience, this cost can reach $1.90 per 1000 gallons or $0.50/m^3 (Asano, 2001). For example, the Sanitation Districts of Los Angeles County in California operate 10 water reclamation plants with a combined capacity of 240.8 mgd (911,428 m^3). In 1997, the districts produced 187.5 mgd (709,688 m^3) of reclaimed water, of which 39.3 percent was actively reused at 371 sites for landscape and agricultural irrigation, industrial process water, recreational impoundments, wildlife habitat maintenance, and groundwater recharge (LACSD, 1998). Table 4.9 compares the rates of potable water with those for reclaimed water in seven retail agencies that use the reclaimed water from the districts in Los Angeles County. The difference in rates does not necessarily represent the production and delivery cost of reclaimed water; instead the purveyors view it as a discount to encourage use.

Table 4.9 Potable vs. Reclaimed Water Rates in Los Angeles County

Water Purveyor	Potable Water $/1000 gallons	Reclaimed Water $/1000 gallons
Long Beach Water Department	2.33	1.16
City of Cerritos	1.27	0.67
City of Lakewood	1.52	1.27
Central Basin WMD	1.42	0.61-0.80
Pomona Water Department	0.88-1.10	0.24-0.37
Walnut Valley Water District	1.92	1.64
San Gabriel Valley Water Company	1.45	0.79-1.23

1000 gallons = 3.785 m^3 Source: LACSD (1999)

4.5.5 DESALINATION

Water desalination will certainly be a factor in the long-term solution to the growing need for freshwater supplies. With 70 percent of the world's population living within 50 miles (80 km) of the oceans, seawater could become an important water supply option. Desalination is not a new technology. In the U.S., desalination of seawater started in Key West, Florida in 1862. Today there are more than 100 desalination plants in the state of Florida (Dziegielewski, 1999). Worldwide, desalination capacity has increased from about 1.5 million m^3/day in 1977 to 26 million m^3/day in 2000, with about 13,600 installations in 120 countries (Semiat, 2000; Abdulrazzak et al., 2002). About 50 percent of the world's capacity is in the Persian Gulf countries with 30 percent in Saudi Arabia. In the U.S., there are about 1,900 installations representing 16 percent of the global capacity. Although the volume of desalinated water is significant, it represents only a small fraction of total water use. For example, in the 13 member countries of the Economic and Social Council of Western Asia (ESCWA), desalination contributes only 1.3 percent of the total water use of 147.3 billion m^3/year (Abdulrazzak, et al., 2002).

The cost of desalination depends on the concentration of salts, plant size, desalination technique, and energy prices. For seawater desalination using various techniques, the unit costs range from $0.75 to $1.5 per m^3 in large plants, and $2 to $3 per m^3 in small plants (Abdulrazzak et al., 2002). The average installation and product costs for several different desalination techniques are compared in Table 4.10. The cost of multi-flush projects is the highest, while costs are similar for reverse osmosis, vapor compression, and multi-effect distillation.

The structure of desalination costs is greatly influenced by the cost of energy. In the Gulf region, the components of total water production cost are influenced by the minimal charges for energy due to subsidies. Table 4.11 shows the approximate structure of water desalination costs in the 11 ESCWA member countries.

Table 4.10 Cost Comparison for Different Water Desalination Techniques

Desalination Technique	Installation Cost ($/m³/day)	Product Cost ($/m³)
Multi-stage flush (MSF)	1,200-1,500	1.10-1.25
MSF (Singapore)	2,300	1.50
Multi-effect distillation (MED)	900-1,000	0.75-0.85
MED (Tower in So. California)	660	0.46
Vapor compression (VC)	950-1,000	0.87-0.95
Reverse osmosis (RO)	700 -900	0.68-0.92
RO (Tampa Bay)	1,000-1,350	0.45-0.56

Source: Semiat (2000)

Energy costs in the Gulf countries account for 20.5 percent of product cost. For comparison, Semiat (2000) reported a typical cost structure of reverse osmosis utilizing seawater as: 44 percent in electric power cost, 37 percent in fixed charges, 7 percent for maintenance and parts, 5 percent for membrane replacements, 4 percent for supervision and labor and 3 percent for other consumables.

Table 4.11 Structure of Desalination Costs in Gulf Countries

Cost Component	Percent of Product Cost
Capital investment	38.0
Energy cost	20.5
Labor	21.3
Maintenance	16.2
Chemicals	4.0

Source: Abdulrazzak et al. (2002)

The cost of desalination is expected to continue decreasing due to technological improvements and other factors. While the theoretical energy requirement to desalinate water is 2 kilowatt-hours per 1000 gallons (0.53 kwh/m^3), the actual energy requirement was 100 kwh/1000 gallons (26.4 kwh/m^3) in 1962. By 1998, energy use in new plants had decreased to 11 kwh/1000 gallons (2.9 kwh/m^3) and many large desalination plants operated at 16 kwh/1000 gallons (4.2 kwh/m^3) (Leitner, 1999). The reduction of energy requirement is largely due to the invention of reverse osmosis. The unit cost of energy can be reduced through the use of dedicated power stations for large desalination plants. Semiat (2000) reported that the cost of electricity from commercial power stations was between $0.06 and $0.07 per kwh because it reflected the need to meet seasonal and daily changes in production. Electricity cost could be reduced to $0.04/kwh if a dedicated power station is used. Finally, the cost of some hardware components used in desalination is also declining. For example, the price of spiral-wound membranes for reverse osmosis declined by more than 60 percent between 1980 and 2000 and will probably continue falling.

There is a common perception among water planners that desalination is prohibitively expensive. In reality, desalination is expensive, but only if compared to the historical cost of developing surface water and groundwater supplies. The development of new sources in many areas may reach a cost of \$0.50 to \$0.75/m^3, which is comparable to the production cost of desalinated seawater. Brackish water with lower salt content can be desalinated at less than half of the cost of seawater.

4.6 Costs in Economic Analysis

4.6.1 ECONOMIC CATEGORIES OF COSTS

In addition to the simple expression of cost as capital investment and O&M or the accounting categories of past expenditures used by water utilities, special categorization of costs is used in economic analysis. The relevant cost categories include fixed and variable, average and incremental (or marginal) as well as short-run and long-run. These categories are important in cost allocation and pricing of water.

The relationship between the total quantity of water delivered during a given time period (or production output) and the total production cost is the basis of the economic evaluation of water supply investments. This relationship is referred to as the total-cost curve. A curve based on time-series data can be developed for a given water utility by summing all monthly expenses (both historical and those anticipated in the future) and plotting them against the monthly quantities of water delivered. Over any given time interval, the total cost can be separated into two parts: fixed costs and variable costs. Fixed costs remain constant over some period of time (i.e., in the short run) and thus do not vary with the quantity of water delivered. They include such expenses as overhead costs and capital recovery charges. Variable costs vary with the quantity of water delivered and include items such as electricity for pumping, chemicals used in water treatment and temporary labor. The total cost (TC), average cost (AC) and marginal cost (MC) functions can be written as:

$$TC = FC + f(W) \tag{4.1}$$

where FC is fixed cost and $f(W)$ is the variable cost function, which depends on the volume of water production W. Average cost is obtained by dividing the total cost by the volume of water production W:

$$AC = \frac{FC}{W} + \frac{f(W)}{W} \tag{4.2}$$

Marginal cost is the mathematical derivative of total cost.

$$MC = \frac{d(TC)}{dW} = f'(W) \qquad (4.3)$$

With a given fixed cost, the average cost curve will be falling because increasing water production spreads fixed costs over a larger volume of water. In the long run, fixed cost will increase if the capacity of the existing system is reached and additional investments in infrastructure have to be made, or it will decrease if the debt for earlier capacity additions is paid off. Because of the presence of economies of scale in water production, capacity is usually added in large "lumpy" increments. Such capacity additions will add new capital recovery charges and in some cases may result in large shifts in the level of fixed cost. On the historic average cost curve, this would appear as a spike followed by a falling trend as the volume of production increases. However, if the new capacity additions coincide with debt retirement, there may not be a significant change in fixed cost.

In economic textbooks, U-shaped average and marginal cost curves are usually depicted, implying that at some level of output the marginal (per unit) cost will start increasing due to diseconomies of scale as shown in Figure 4.4.

4.6.2 EXAMPLES OF AVERAGE COST CURVES

In the case of water supply, the long-run average cost curve will be increasing if the cost of new increments to system capacity is increasing. However, there is extensive evidence that the supply capacity cost functions exhibit economies of scale, which implies that long-term marginal cost will be falling and will plot below the average cost curve. Some evidence of this can be seen in Tables 4.3 and 4.4, which show the declining average total expense per unit as the size of water supply system increases. The values of total expenses in \$/1000 gallons (which include depreciation and interest payments but exclude some capital payments) are plotted against average production in million gallons per day on Figure 4.5.

The average cost curves for surface water and groundwater systems plotted in Figure 4.5, while being only rough approximation of the cost relationships in the water industry, do show a declining trend over the range of water production up to 35 mgd which is 132,475 m^3/day (or approximate population served of 200,000). In surface water systems, average costs declines very rapidly in the range between 0 and 1.5 mgd (5,678 m^3) and remains relatively constant between 1.5 and 32.5 mgd. However, it is important to note that the data in Table 4.4 and Figure 4.5 are derived from a number of observations of different individual plants and do not apply to any particular site. Although these data do not indicate an increasing average cost of water supply in the water industry, they do not constitute a proof that the water industry is facing a constant or declining marginal cost in the long run. Some individual systems have "planned" capacity expansion paths that clearly suggest an increasing marginal capacity cost. Some examples of marginal costs are presented in the next section.

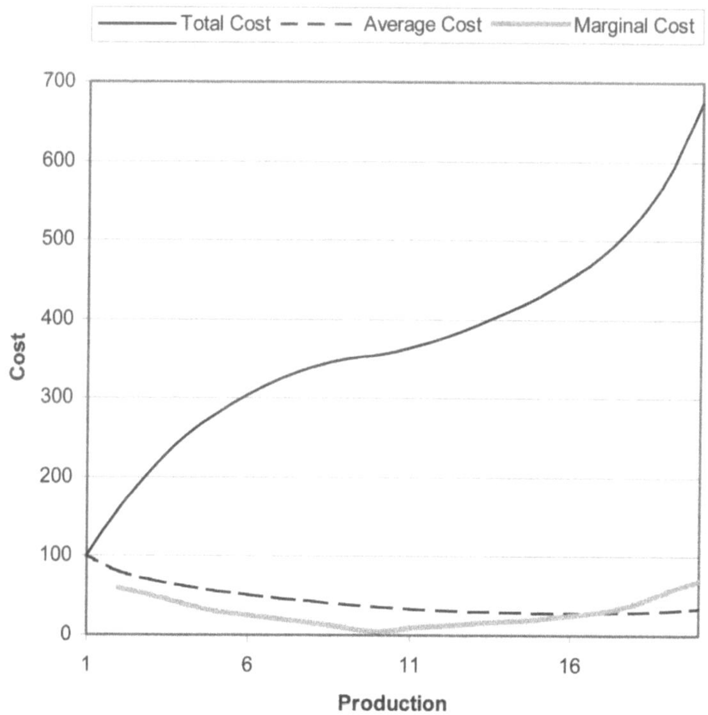

Figure 4.4 Total, Average and Marginal Cost Curves

4.6.3 MARGINAL CAPACITY COST

Economists are interested in the long-term marginal cost of water supply because economic theory suggests that setting water price at the level of marginal cost will induce consumers to use water efficiently. However, in order to put this theory into practice, it is necessary to determine the actual value of marginal cost of water supply.

Calculation of marginal cost of water supply requires information on the cost of plant and equipment as well as future investment needs. The cost of water supply can be separated into several components based on the nature of water supply operations and the physical components of the systems. The total cost can be separated into the cost of raw water (source, intake and transmission), cost of water treatment (including pumping), cost of distribution and storage, and customer cost (including metering and billing). For each component, total cost can be separated into capital (fixed) cost and variable cost. Also, average cost and marginal cost can be calculated. It is possible that marginal cost will be declining with the level of production for some components and rising for others.

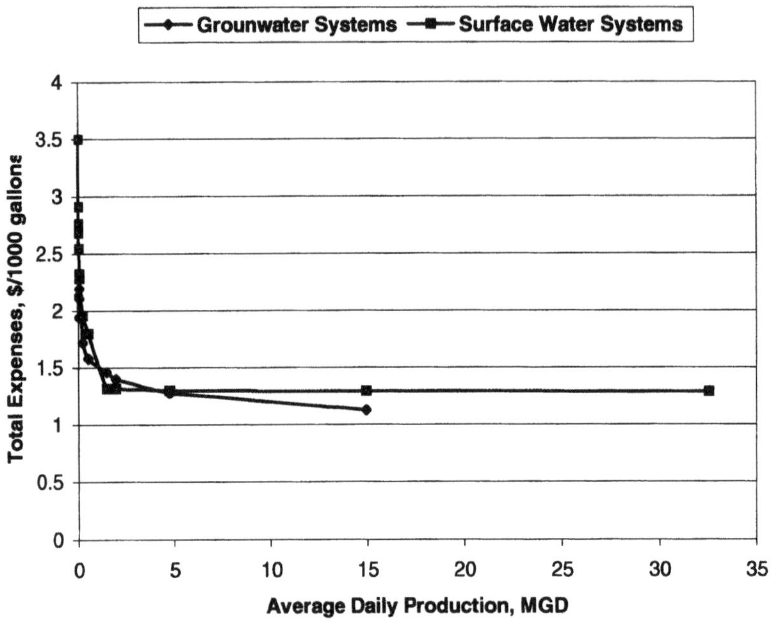

Figure 4.5 Approximate Average Cost Curves for Water Supply Systems (Derived from EPA survey data, see Table 4.4)

One concern commonly found in the literature is that due to increasing scarcity of water, the cost of developing additional water supplies will inevitably rise. This implies that the marginal cost of one component of water supply systems, namely the capacity cost of water supply sources will be increasing. However, at the national level, the need for new sources represents a relatively small part of the total investments for drinking water infrastructure. According to the EPA, the total 20-year need for new sources (including construction and rehabilitation of surface water intake structures, raw water pumping facilities, drilled wells and spring collectors) is $9,599.6 million or 6.4 percent of the total investment (see Table 4.12). This suggests that even if the unit cost of developing new sources of supply is rising, it likely will be concealed by the costs of transmission and distribution and water treatment facilities.

4.7 Capacity Expansion Costs in Phoenix

Unfortunately, no national data on the costs of new additions to supply or treatment capacity at specific sites or systems are available. One example of the future costs of raw water supply and treatment capacity comes from a study of the Phoenix Water

Services Department (WSD) that was conducted by Russell and Shin (1993). The study analyzed marginal cost of source capacity based on the actual capacity expansion plans for the period 1993-2010. Also included was a consideration of different marginal costs during high-use and low-use seasons (or peak and off-peak costs).

Table 4.12 Total Cost of New Drinking Water Infrastructure

Investment Purpose	Estimated 20-year Need (millions of 1999 dollars)	Percent of Total Need
Transmission and distribution	83,154.2	55.1
Water treatment	37,962.2	25.1
System storage	18,365.8	12.2
Water sources	9,599.6	6.4
Other	1,857.5	1.2
Total	150,939.4	100.0

Source: EPA (2001)

The planned raw water supply projects included acquisition of new water rights and agricultural water transfers, expansion of reservoir storage capacity, groundwater recharge projects, construction of new wells and water reclamation. Figure 4.6 shows three plotted cost lines: one represents total source capacity in million hcf (hundred cubic feet), another shows accumulated new annual investment costs, and the third shows the incremental cost of new capacity in dollars per 10 hcf/year. The incremental cost line shows relatively costly additions to capacity before 1998 and after 2005. The low unit cost additions represent new wells with an estimated cost of $500,000 per well yielding 1,757,000 hcf (4,972,000 m^3)/year or $0.285/hcf /year ($0.101/m^3/year).

In addition to source capacity, Phoenix WSD also developed plans for additions to water treatment capacity. The capacity and costs of these additions are shown in Table 4.13. The unit cost of these planned additions varies between $5.3 and $13.2 per cubic-foot/day ($709 to $1,765 per 1000 gallons/day or $187 to 466 per m^3/day). The lowest costs are for two plants to be completed in 2005.

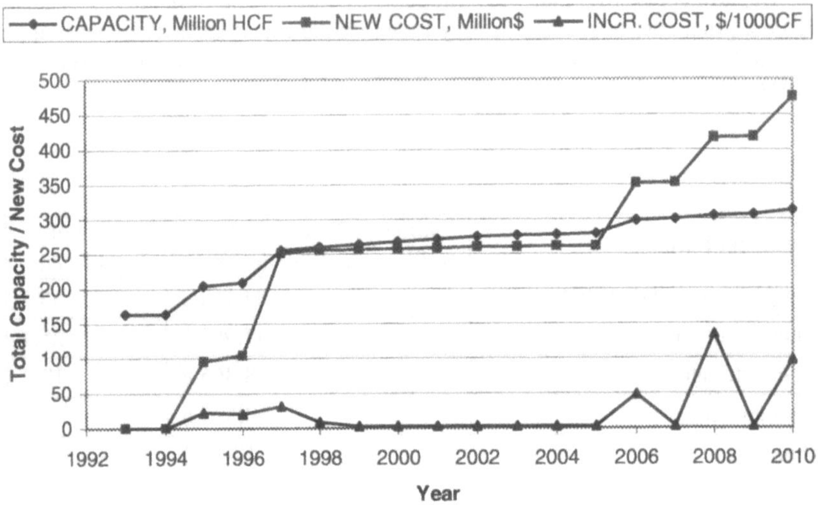

Figure 4.6 Planned Raw Water Capacity Additions and Estimated Costs for Phoenix, Arizona derived from data in Russell and Shin, 1993.

Table 4.13 Plant Additions to Water Treatment Capacity in Phoenix

Year Completed	Name of Treatment Plant	Capacity (1000 CF/day)	Investment Cost, $1000	Unit Cost $/CF/day
1993	Existing plants	66,845	--	--
1995	Verde	2,674	30,000	11.2
1995	Deer Valley	4,011	53,000	13.2
1997	Val Vista	5,348	34,000	6.4
2000	Squaw Peak 1	5,348	54,000	10.1
2005	Squaw Peak 2	4,011	15,000	3.7
2005	Lake Pleasant	10,696	57,000	5.3

Source: Derived from data in Russell and Shin, 1993

Russell and Shin (1993) used the information on the planned capacity investments in Phoenix to calculate marginal cost prices for peak and off-peak seasons. They compared three alternative methods for approximating marginal cost prices: (1) "textbook method" in which marginal price is equated to the average (per unit) annualized cost of the capacity increment, (2) Turvey method, in which the marginal cost of a single capacity increment is the cost of having to do it one year earlier, and (3) a modified "average marginal cost method" proposed by Saunders et al. (1977). The resultant marginal cost prices (which include the marginal capacity cost and average variable cost) for each method of approximation are given in Table 4.14. Figure 4.7 shows a plot of marginal cost prices for peak season obtained through each method.

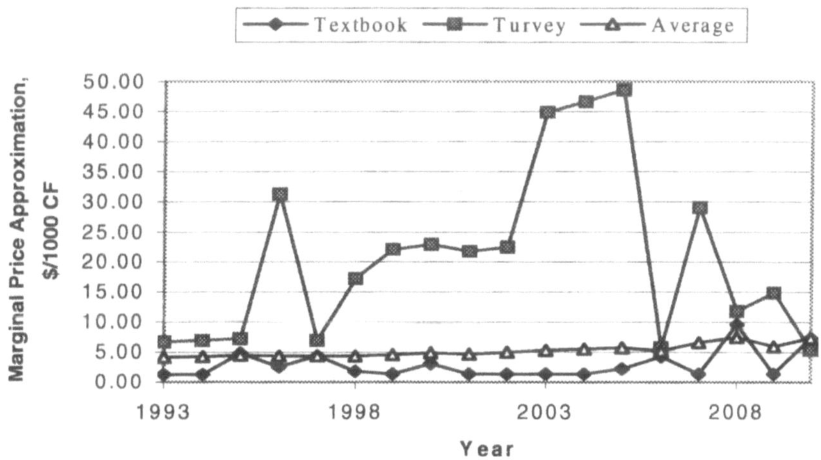

Figure 4.7 Marginal Cost Prices for Peak Season in Phoenix.
Source: Derived from Russell and Shin, 1993.

The Turvey method resulted in highly variable prices during the 18-year period. The most stable marginal cost-based price was obtained using the average marginal cost method. The peak-season marginal price gradually increases from the 1993 value of $2.90/1000 cubic feet to $7.30/1000 cubic feet in 2010. For comparison, the actual marginal price in Phoenix in 1993-1994 period ranged from $0.79 to $1.24 per hundred cubic feet (HCF) depending on season (Kiefer, 1994; Willett, 1999). After converting these rates into units used in Table 4.14, the actual marginal prices ranged from $7.90 to $12.40 per 1000 cubic feet ($0.28 to $0.44/m^3).

Russell and Shin (1993) concluded that if the Phoenix WSD replaced its then current pricing scheme with the average marginal cost price, it could realize efficiency gains on the order of 10 percent during the 18-year planning horizon without excessive price volatility or revenue fluctuations.

Table 4.14 Peak and Off-peak Marginal Cost Prices for a Planned Capacity Expansion Path in Phoenix

Year	Off-peak Season Marginal Price* $/1000 cubic feet			Peak Season Marginal Price* $/1000 cubic feet		
	Textbook	Turvey	Average	Textbook	Turvey	Average
1993	1.24	4.75	2.90	1.24	6.69	4.15
1994	1.24	4.91	2.98	1.24	6.94	4.29
1995	2.67	5.07	3.06	4.78	7.19	4.43
1996	2.52	29.53	3.25	2.52	31.31	4.30
1997	3.22	3.94	3.29	4.30	6.85	4.39
1998	1.79	14.93	3.36	1.79	17.18	4.38
1999	1.41	19.78	3.55	1.41	22.13	4.61
2000	1.41	20.50	3.75	3.12	22.96	4.86
2001	1.41	21.23	3.99	1.41	21.72	4.69
2002	1.41	22.00	4.26	1.41	22.51	4.99
2003	1.41	44.38	4.58	1.41	44.91	5.34
2004	1.41	46.20	4.76	1.41	46.76	5.55
2005	1.41	48.08	4.95	2.24	48.66	5.78
2006	4.20	5.76	5.15	4.20	5.76	5.15
2007	1.41	29.08	6.60	1.41	29.08	6.60
2008	9.60	11.82	7.40	9.60	11.82	7.40
2009	1.41	14.91	5.91	1.41	14.91	5.91
2010	7.30	5.45	7.30	7.30	5.45	7.30

* Price estimation for each method of approximation is based on a 4.5 percent discount rate

Source: Russell and Shin (1993)

4.8 Summary

This chapter has examined the costs of urban water supply using existing data from various publications and EPA surveys as well as examples of site-specific data from Phoenix. The existing data and information support three general conclusions.

First, there is an increasing interest in the issues surrounding the concept of the value of water in alternative uses and its relationship to the principle of sustainable management and use of water resources. Under the sustainability criteria, the full cost of water use (including the cost of water supply, opportunity cost and both economic and environmental externalities) should equal the full value in use. A few cases studies cited here indicate that the full cost of agricultural and urban uses is greater than the full value derived by these uses. Only industrial uses appear to derive values that exceed the full cost. Also, the prices paid by water users are significantly below the actual full cost of water supply and far below the economic cost.

Second, urban water supply agencies have a range of conventional and unconventional options for developing additional sources of water supply. Management of current and future demand is a relatively inexpensive alternative for reducing water demand and can be viewed as a viable source of water. Two other alternatives, water reclamation and desalination are more expensive than traditional surface water or groundwater supplies but the difference in costs is narrowing and their widespread use is very likely in the coming decades.

Third, many water supply systems face increasing long-term marginal cost of water supply. However, the near-term cost increases appear to be moderate and are related more to the increasing costs of water treatment to meet the SDWA standards than to the high costs of new sources of supply. Furthermore, marginal cost-based water pricing schemes can lessen the need for developing new supply sources when consumers react to the "true" long-run marginal prices by reducing water demand.

In light of these conclusions, future research in the area of water supply economics could focus on better documentation of the actual cost of water supply, especially the cost of unconventional alternatives, and on development of pricing schemes that not only promote cost responsibility pricing but also provide consumers with information on the full cost of water use. The later represents an important prerequisite for achieving sustainable use of water resources which is much advocated today.

References

Abdulrazzak, M.J., M. Jurdi, and S. Basma, 2002. "The Role of Desalination in Meeting Water Supply Demands in Western Asia." *Water International* 27(3).

Asano, T., 2001. Water from Wastewater: The Dependable Water Resource. Paper presented at Stockholm Water Symposium, August 12-18, 2001, Stockholm, Sweden.

Boisvert, R.N. and T.M Schmidt, 1996. *Economies of Size in Water Treatment vs. Diseconomies of Dispersion for Small Public Water Systems.* Working Paper 96-15. Department of Agricultural, Resource and Managerial Economics, Cornell University.

Dziegielewski, B. and D.D. Baumann, 1992. "Benefits of Urban Water Conservation." *Environment* 34(9): 7-41.

Dziegielewski, B., E.M. Opitz, J.C. Kiefer, and D.D. Bauman, 1993. *Evaluating Urban Water Conservation Programs: A Procedures Manual,* American Water Works Association. Denver, Colorado.

Dziegielewski B. and W.Y. Davis, 1997. *Reassessment of Water Conservation.* Consultant report prepared for Eugene Water and Electric Board. Eugene, Oregon.

Dziegielewski, B., 1999. "The Role of Desalination in Averting a Global Water Crisis." Conference Report. *Water International* 24 (4): 395-398

Frederick, K.D., 1995. "America's Water Supply: Status and Prospects for the Future." *Consequences* 1(1).

Herman, S.W., 2000. "Fixed Assets and Consumer Durable Goods: Estimates for 1925-98." *Survey of Current Business.* April.

Howe, C. and S. White, 1999. "Integrated Resource Planning for Water and Wastewater: Sydney Case Studies." *Water International* 24(4): 356-362.

Hoffman, M., R. Glickstein, and S. Liroff, 1979. "Urban Drought in the San Francisco Bay Area: A Study of Institutional and Social Resiliency." *Journal, American Water Works Association,* 71(7): 356-363.

Kiefer, J.C. 1994. *City of Phoenix Water Use Monitoring Program: A Multiobjective Study of Single-family Household Water Use.* Technical Report, Planning and Management Consultants, Ltd., Carbondale, Illinois, January.

Los Angeles County Sanitary District, 1999. *Water Reuse Summary for Fiscal Year 1996-97.* http://www.lacsd.org/wastewater/.

Lower Colorado Regional Water Planning Group, 2000. *Groundwater Supply Alternatives for the Cities of Blanco, Goldthwaite and Llano.* Technical Memorandum, Prepared by CH2M Hill. December.

Leitner, G. 1999. Conference Presentation, Reported in Dziegielewski (1999).

Rogers, P.R. Bhatia and A. Huber, 1998. *Water as a Social and Economic Good: How to Put the Principle into Practice.* Global Water Partnership, Stockholm.

Russell, C.S. and B. Shin, 1993. *Marginal Cost Pricing Rules for System Net Benefits, Demand Behavior, and Water Utility Pricing: Conceptual Underpinnings and Case Study of the Phoenix Water and Wastewater Department.* USGS Grant Report 14-08-0001-G2086. Planning and Management Consultants, Ltd., Carbondale, Illinois.

Saunders, R.J., J.J. Warford and P.C. Mann, 1997. *Alternative Concepts of Marginal Cost for Public Utility Pricing: Problems and Applications in the Water Supply Sector.* World Bank Staff Working Paper No. 259, Washington, D.C.

Solley, W.B., R.R. Pierce and H.A. Perlman, 1998. *Estimated water Use in the United States in 1995.* U.S. Geological Survey Circular 1200, Reston, Virginia.

Semiat, R., 2000. "Desalination: Present and Future." *Water International* 25(1):54-65.

US Army Corps of Engineers, 1998. *Water Supply Handbook.* IWR Report 96-PS-4. Institute for Water Resources, U.S., Alexandria, Virginia.

U.S. Bureau of Economic Analysis, 2000. *Fixed Assets and Consumer Durable Goods in the United States, 1925-1997.*

U.S. Environmental Protection Agency, 1997. *1996 Clean Water Needs Survey.* Cincinnati, Ohio.

U.S. Environmental Protection Agency, 2001. *Drinking Water Infrastructure Needs Survey: Second Report to Congress,* EPA 816-R-01-004, February.

Willett, J.S., 1999. Personal Communication. Phoenix Water Services Department.

Warford, J.J., 1997. *Marginal Opportunity Cost Pricing for Municipal Water Supply.* Special Paper, International Development Research Center. Ottawa, Canada.

Chapter 5

Elasticity of Demand for Water Resource Managers

Donald E. Agthe
Economic Consultant
R. Bruce Billings
University of Arizona

5.1 Introduction

Demand management policies for water systems should be based on knowledge of economic behavior of the water customers. Some aspects of this economic behavior can be quantified using the concept of elasticity. Relatively simple calculations can guide the water system management in solving questions of equity and efficiency in water production and distribution. While the concept of elasticity can be applied to an almost endless number of topics, this chapter discusses the utilization and calculation of price elasticity of demand, cross price elasticity of demand, and income elasticity of demand for water.

5.2 The Concept of Elasticity

Elasticity is defined as the relative response of one variable to the change in another variable. Thus, it is useful in describing the response of a particular variable to a change in policy or design associated with a second but related variable. The elasticity coefficient is usually stated as the percentage change in variable Y associated with a one percent change in variable X. Equation 5.1 is the most intuitive definition of elasticity.

$$\text{Elasticity (of X with respect to Y)} = \frac{\text{Percent Change in X}}{\text{Percent Change in Y}} \tag{5.1}$$

D.E. Agthe et al. (eds.), Managing Urban Water Supply, 71–86.

5.3 Price Elasticity of Demand

Price elasticity of demand is used to measure the relative change in quantity of water demanded that results from a one percent change in the price of water. This concept is often referred to as *own price elasticity of demand,* as defined by equation 5.2.

$$\text{Price Elasticity} = \frac{\text{Percent Change in Quantity}}{\text{Percent Change in Price}} \tag{5.2}$$

The price used in this computation could be either an average or marginal price. Average price is obtained by dividing water utility revenue by total quantity delivered. The price could also be computed in this way for each customer class, or for individual customers. Revenue may include fixed periodic charges, commodity charges, and sewer use fees. Marginal price is the price of the last unit of water purchased. For example, if the utility has a single rate for all water, that rate is the marginal price. For an increasing block rate structure, it is the highest block in which a customer buys water. The effect of changes in block rates are not readily computed with the simple elasticity equation since customers will be in different blocks. The results of the elasticity computation will differ between using marginal and average price.

It is important to understand what is being measured when policy decisions are being made based on an elasticity coefficient. Given that the demand relationship shows the quantities that will be taken from the market associated with various prices, price elasticity of demand measures the change in *quantity demanded* in response to a change in price rather than a change in demand itself. A change in demand requires a shift of the demand curve due to the change in a non-water price influence. The other two measures of elasticity of demand presented in this chapter, cross-price elasticity and income elasticity, are associated with the shape and location of the demand curve. Therefore they measure factors that change demand. These factors are called "demand shifters."

Price elasticity of demand between two points on a demand curve can be calculated using the simple equation 5.3 or the midpoint equation 5.4.

$$\text{Price Elasticity of Demand} = \frac{\dfrac{\text{Quantity A - Quantity B}}{\text{Quantity A}}}{\dfrac{\text{Price A - Price B}}{\text{Price A}}} \tag{5.3}$$

$$\text{Price Elasticity of Demand} = \frac{\dfrac{\text{Quantity A - Quantity B}}{\text{QuantityA + Quantity B}}}{\dfrac{\text{Price A - Price B}}{\text{Price A + Price B}}} \quad (5.4)$$

The midpoint equation will provide more accurate forecasts than the simple elasticity calculation shown in equation 5.3 because it yields the same coefficient whether the price rises or falls over the segment of the demand curve being measured. For example, suppose at a price of \$2.00 a group of customers buys 4,500 units of water per month but when the price is increased to \$3.00 they buy 3,700 units. Using equation 5.3 the elasticity would be computed as:

$$\text{Price Elasticity of Demand} = \frac{\dfrac{4500 - 3700}{4500}}{\dfrac{\$2.00 - \$3.00}{\$2.00}} = -0.36 \quad (5.5)$$

However, if price had been reduced and the quantity increased, 5.3 would give the following result.

$$\text{Price Elasticity of Demand} = \frac{\dfrac{4500 - 3700}{3700}}{\dfrac{\$2.00 - \$3.00}{\$3.00}} = -0.65 \quad (5.6)$$

In contrast, using equation 5.4 yields a result which is useful for both price increases and decreases, and will provide reasonably accurate predictions when prices change.

$$\text{Price Elasticity of Demand} = \frac{\dfrac{4500 - 3700}{4500 + 3700}}{\dfrac{\$2.00 - \$3.00}{\$2.00 + \$3.00}} = -0.49 \quad (5.7)$$

Unlike these simple elasticity computations, which ignore the effects of other influences on demand such as the weather and income changes, econometric estimation techniques discussed at the end of this chapter can take many influences into account.

A price elasticity of demand of -1.0 is unitary elastic while price elasticities of demand of less than minus one (ie. larger negative numbers) are price elastic and those greater than -1 (between -1 and 0) are inelastic. Since the demand curve is almost always negatively sloped, economists regard positive price elasticities of demand skeptically and tend to dismiss them as unrealistic. Houthakker and Taylor (1970) suggest that extremely low prices, occasionally the case for water, may lead to estimates of price elasticities of demand greater than zero. This is a result found in an urban water demand study of Chicago by Wong (1972). Examples of calculated price elasticities of demand for water are shown in Table 5.1, which shows a range of price elasticity estimates, varying with geographic region, season, use category and income level.

Table 5.1 Examples of Price Elasticities of Demand for Water

Region or Industry	Subclass	Annual	Winter	Summer
Tucson, Arizona[1]	Low Income	-0.51		
	Medium Income	-0.39		
	High Income	-0.57		
Texas[2]	Seasonal		-0.19	-0.37
Oakland, Califronia[3]	Seasonal		-0.10	-0.17
Toronto, Ontario[4]	Seasonal		-0.75	-1.07
Cincinnati, Ohio[5]		-0.62		
Honolulu, Hawaii[6]		-0.35		
Kansas[7]		-0.69		
Stone,Clay[8]	Industrial	-0.38		
Metal Fabrication[8]	Industrial	-0.41		
Petrochemicals[9]	Industrial	-0.12		
Forest Industry[9]	Industrial	-0.51		

[1]Agthe and Billings (1997); [2]Griffin and Chang (1990); [3]Weber (1989) ;[4]Grima (1972); [5]Clark and Goddard (1997); [6]Muncur (1987); [7]Gottlieb (1963); [8]Babin (et al., 1981); [9]Rensetti (1988)

The calculated coefficient for price elasticity of demand depends heavily upon a number of factors:

- Percent of consumer budget spent for water
- Price of water
- Price of substitutes for water (e.g. plumbing repairs, dirty cars)
- Number and closeness of substitutes
- Number of uses
- Role of water in the production of a commercial good
- Consumer demand for goods and services using water

A price elastic good is often a good with many substitutes that are relatively close in price, a relatively high price that makes it an important part of the consumer or producer budget, and/or a large number of actual or potential uses. These criteria are often interrelated. For example, a good with many uses probably has several close substitutes in some of its uses and, therefore, is likely to be price elastic. Price inelastic goods tend to have the opposite set of characteristics. They have few substitutes close in price, are relatively small in the consumer budget, and have lower prices. Thus, price elastic consumer goods tend to be luxury goods while price inelastic goods tend to be necessities. In most cases drinking water supplied by the local water company is price inelastic because of its low price, lack of close substitutes, and small importance in consumer budgets.

There may be situations when water is price elastic. Bottled water sold by specialized companies is likely to be price elastic as it is sold at a much higher price per gallon than city water. Sometimes the end use can be a luxury use bringing about an elastic demand for water. For example, backyard swimming pools, golf courses, and fine gardens may be price elastic markets for urban water when the price of water is sufficiently high.

5.4 Derived Demand for Water as a Factor of Production

The demand for water used as a factor of production for a good or service is called derived demand. Most water that is not used for human consumption is derived from the demand for other final goods and services. Thus demand for water is determined by both the demand for the final product and the role of water in producing the good. The demand derived from the production relationship and final product can be price elastic if utility supplied water has close substitutes (including the use of private wells) in the production process. Close substitutes in production are important because a shift in the price of water relative to its substitutes may change the demand for water. If the price of a close substitute in production rises, water may be substituted for that input. On the other hand, if the price of water rises then less water will be used and more of the substitute input will be used. Examples of derived demand for water include water used in production of pleasant landscapes, games of golf, bottled sodas, and industrial process cooling.

5.5 Price Elasticity and Revenue

The price elasticity of demand is directly related to the extra revenue that can be obtained from a price increase. Equation 5.8 shows the relationship between the price elasticity of demand and the corresponding price elasticity of revenue. Price elasticity of revenue is the percentage change in revenue for a one percent change in the price of water.

$$\text{Price Elasticity of Revenue} = \text{Price Elasticity of Demand} + 1.0 \qquad (5.8)$$

This simple computation avoids a common pitfall of assuming that a 10% price increase will increase revenues by 10%. Such a result is highly unlikely! For example, if the price elasticity of demand is -0.4, the corresponding price elasticity of revenue is +0.6, so that a 10% price rise results in quantity dropping by 4% and total revenue rising by 6%. As long as price elasticity of demand is between 0 and -1.0, a price increase will increase revenue. When the price elasticity of demand is -1.0, the price elasticity of revenue is 0.0! In this case, a price rise causes a proportional drop is sales and no change in revenue. Although revenue is constant in this example, the price rise may still be helpful in reducing an operating deficit, since the resulting drop in quantity will result in a somewhat lower total cost of operation.

Figures 5.1a and 5.1b relate changes in quantity demanded and revenue received to the price elasticity of demand found along a linear demand curve. The elasticity varies continually along the demand curve even though the slope of the curve remains constant. This is because slope shows the absolute change in quantity for any price change while elasticity shows the relative change. Figure 5.1b shows that the maximum possible revenue is received at the midpoint of the quantity range of the linear demand curve which is also the midpoint of the total revenue curve. Demand is unitary elastic at this midpoint. At smaller quantities on these curves, demand is relatively elastic and the price of water must be decreased to increase revenue. Thus, a water manager facing an elastic demand and revenues that do not cover costs can only solve this problem by raising price if the resulting drop in volume reduces cost by an even larger amount than the decrease in revenue. This is unlikely, so such a manager would usually be better served by lowering the price and selling more water and/or by finding ways to reduce costs by increasing operating efficiency. At quantities larger then the midpoint of these curves, demand is inelastic. With inelastic demand, which is typical for water utilities, a water manager can increase revenue by increasing price. Since low income customers tend to have more inelastic demands than higher income customers with more luxury uses for water, the water manager may have to be careful to maintain equity in the price rise scheme.

5.6 Uses of Price Elasticity of Demand

Knowledge of the price elasticity of demand can assist the water manager in forecasting the changes in sales volume and resulting revenue resulting from a proposed price change. For example, if the price elasticity of demand is -0.36, a 9 percent price hike could be expected to reduce volume sales by 3.24 percent and increase revenues by 5.76 percent. Estimates of price elasticity for each customer class are useful in setting rates which vary by customer classes. Industrial and recreational markets are likely to have different responses to price changes than residential markets. Even within residential markets, consumers with different income levels may have different elasticities (Agthe and Billings, 1987). For equity reasons, it may be necessary to offer low-income customers a lower water rate or to develop a special program to suit their needs when a price increase is imposed. Different parts of a city may have different price elasticities because of lot sizes or some other housing characteristic. For example, individually metered condominium owners, cooperative owners and apartment renters are likely to

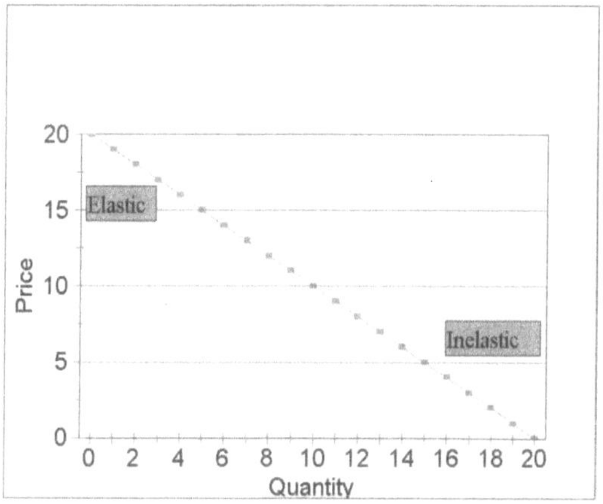

Figure 5.1a Linear Demand Curve

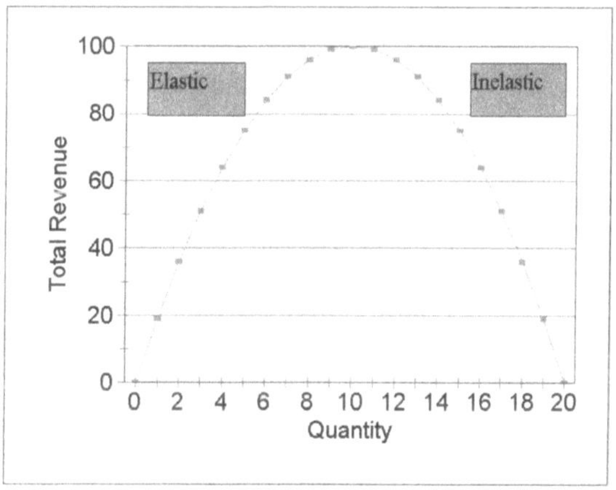

Figure 5.1b Corresponding Total Revenue

use almost all indoor water while individual residences frequently have substantial garden and lawn areas. Because of their relatively short tenure compared to condominium and cooperative owners, renters have little incentive to install water saving toilets and plumbing fixtures. While most utilities charge uniform rates within their service areas, large differences by area in elasticities may be useful in planning for changes in delivery capacity. Water managers must be aware of the price elasticities in their submarkets to effectively manage revenues and prevent public trust from being eroded by inequitable practices.

Table 5.1 shows that summer price elasticities of demand for cities are typically larger in summer in North America. This suggests that water has more discretionary uses and more substitutes in the summer. It also suggests that education programs for outdoor water conservation, higher summer rates, and higher rate blocks for summer use may be more effective means of conserving water than programs aimed at winter and indoor water use. Price can be an effective conservation tool as long as prices are increased periodically to keep up with the inflation rate of the general economy.

We suggest as a rule of thumb that charging prices that make the price elasticities of demand equal across markets would perhaps be the most equitable arrangement of prices when a water manager is selling to different end use markets with a wide range of customer incomes. If prices are set in this way, each market would cut water use by the same percentage in response to a proportional price increase.

5.7 Cross-Price Elasticity of Demand

The cross price elasticity of demand is a measure of the ability of some other good or service to complement water or substitute for water. The cross price elasticity of demand for good "A" with respect to a change in the price of water is shown in equation 5.9.

$$\text{Cross Price Elasticity} = \frac{\text{Percent Change in Quantity of Good A}}{\text{Percent Change in Price of Water}} \tag{5.9}$$

An example of water being a complementary good is the filling of a swimming pool. If the price of water is low, the building of swimming pools is encouraged but if the price of water is sufficiently high, the building of swimming pools is discouraged. Examples of substitutes for water include the use of low flow plumbing devices and the frequency of leak repairs. If the price of water increases, plumbing repair is encouraged and homeowners are encouraged to convert their plumbing to low flow devices. Thus, increasing the price of water relative to the price of substitutes for water can be an effective conservation tool.

A negative cross price elasticity coefficient indicates a complementary relationship and a positive coefficient indicates a substitute relationship. For example, if the cross price elasticity of demand for plumbing repairs with respect to the price of water is +0.23, a 10 percent increase in water rates would result in an increase of 2.3 percent in spending on repairs. In this example, the effect is likely to occur soon after the higher water bill is received. For a different example, if the cross price elasticity of demand for acres of grass on golf courses with respect to the price of water was -0.34, a 10 percent increase in water rates would result in a 3.4 percent reduction in grass covered areas. It might be a number of months before golf course managers would make the predicted changes. In both examples, some people would respond quickly to the change in relative prices while others would respond slowly or not at all. Elasticity estimates are based on the average or typical behavior of large groups of people and not on particular individuals. Thus the observation that some people ignore a price change and make no response does not invalidate an elasticity based forecast.

Cross price elasticity can also be used to describe the effect of a change in the price of a complement or substitute on water use. The cross price elasticity of demand for water with respect to the price of good "A" is:

$$\text{Cross Price Elasticity} = \frac{\text{Percent Change in Quantity Water}}{\text{Percent Change in Price of Good A}} \qquad (5.10)$$

Knowledge of the cross price elasticity of demand gives water managers additional options for conservation management. A large cross price elasticity for water with respect to the price of low flush retrofit toilets might suggest instituting a rebate for their installation as an effective conservation measure. Some cities have offered devices to reduce toilet flush volume and low flow showerheads free of charge (thereby lowering their installed cost to homeowners) to encourage water conservation.[13] The effectiveness of these types of conservation programs is discussed in Morgan and Pelosi (1980) and Witcomb (1990). Rebates have also been given for replacing older 5 and 7 gallon toilets with new 1.5 gallon (5.7 liter) units. Similarly, incentives to install low water use landscapes in desert areas could be effective if the cross price elasticity of water with respect to landscaping cost is high. Some cities in Arizona and elsewhere have used such programs. Unfortunately, few studies have been completed evaluating the effectiveness of these programs.

5.8 Income Elasticity of Demand

As consumer incomes increase they tend to increase their consumption of most goods, including water, but at a slower rate than the increases in income. Income elasticity of demand for water is defined as the percentage change in the quantity of water purchased resulting from a one percent change in income.

$$\text{Income Elasticity} = \frac{\text{Percent Change in Quantity of Water}}{\text{Percent Change in Income}} \qquad (5.11)$$

Numerous studies have estimated an income elasticity of demand between 0.2 and 0.4. This suggests that water use generally increases by about 3 percent for every 10 percent increase in income. The reason for this increase is that water using features such as larger lawns, additional bathrooms, decorative fountains, and swimming pools are added to the household as income rises. More frequent laundering of clothes and washing of vehicles is also likely.

The income elasticity coefficient could be greater than one in some unusual situations. For example, a sudden increase in local incomes resulting from a new factory or government facility could encourage widespread adoption of the previously mentioned water using devices combined with new golf courses, water parks, and pools.

[13] The cost is generally not zero, since the homeowner must still install the device or at least take time to allow a city worker to make the installation.

However, the large response to this income increase is likely to be more of a one-time event than a continuing phenomenon. The income elasticity might also be greater than one if rising income allow households to hook up to the piped water system to replace water previously brought to the home by hand or delivery truck.

Income elasticity of demand is important to urban water supply planners. If a new industrial, government, or commercial facility is located in the community, it becomes important to calculate the expected increase in residential water use resulting from the higher incomes in addition to the expected increase in residential water use resulting from any population growth associated with the facility. If a particular segment of the local population is to benefit more economically than the other segments, the income elasticity of water demand for this segment should be known so that water use increases can be accurately predicted.

5.9 Economic Modeling to Estimate Elasticity

The price elasticity of demand can be computed using the midpoint equation or even simple percentage changes. However, if other variables changed significantly during the time period being examined, a comprehensive demand model utilizing several independent variables is needed to determine the effect of the price change. For example, if price is increased at the beginning of an unusually wet summer season, the weather and price both act to reduce water use. A substantial increase in community income and its accompanying increase in water use may appear to make a price increase ineffective. There are several good sources of detailed information on demand modeling, including Billings and Jones (1996).

5.9.1 TIME SERIES MODELS

Time series data are often used in estimating the response of water use to changes in price and other variables. A typical time series analysis uses annual or monthly data, although other periodic data can be used, depending on the objective. We suggest a minimum of four years of monthly data or twenty years of annual data to obtain valid results. A typical economic model (5.10) for urban water demand uses quantity of water as the dependent variable and a number of variables such as weather, price of water, and income as influence (independent) variables.

$$Q = f(P, Y, W) \tag{5.12}$$

Q = quantity of water	P = price of water
Y = income	W = weather variable(s)

For a model using aggregate monthly data, Q is total water deliveries per month, obtained from the water utility divided by population (or number of connections). Matching price data can be obtained from utility records. Average price can be computed as total system revenue from operations divided by total water deliveries.

Community income per capita (or per household) may be available from the Department of Commerce, or a state or local economic development agency. Given a choice, median income (the income of the person in the middle of the distribution) is preferred to average income, since the average may be distorted by extreme values (e.g. Bill Gates lives in the area). However, when median income is unavailable, average income is a good substitute. Monthly income values often must be interpolated from available annual statistics.

In all time series work, the nominal economic variables such as price and income must be adjusted for inflation to create real price and income estimates. With constant (nominal) water rates and four percent per year inflation, real prices are dropping 4%/year, resulting in likely rising water use. If it exists, a local or regional consumer price index would be used to deflate these monetary values. When local data are not available, a national consumer price index or GDP deflator can be used. National deflators are usually available from the government, the central banking authorities, or an international agency. The equation for deflation of an economic series is:

$$\text{Real Value} = \frac{\text{Nominal Value} \times 100}{\text{Deflator Index}} \tag{5.13}$$

For example, if the GDP deflator for January is 125.8 and January income is $2,200, the inflation-adjusted January income in real (inflation adjusted) dollars is $1,748.81.

$$\text{Real January Income} = \frac{\$2{,}200 \times 100}{125.8} = 1{,}748.81 \tag{5.14}$$

Weather can be measured in a number of ways. Researching in an urban desert environment, Billings and Agthe (1980) chose rainfall minus the evapotranspiration of Bermuda grass, the most common ground cover, as their weather variable. Howe and Linaweaver (1967) utilized the Thornthwaite (Thornthwaite and Mather, 1955) method of estimating potential evapotranspiration in their classic national study of residential customers. Estimates of evapotranspiration can be computed from readily available weather data. Simpler variables have sometimes generated more accurate results than evapotranspiration (Billings and Jones, 1996). Some studies include separate variables for temperature and rainfall (Agthe and Billings, 1997). Others have utilized temperature, rainfall, and a vector of monthly dummy variables (omitting one month as the reference level) to capture systematic seasonal variation (Weber, 1989). Temperature and rainfall in this case may be entered as deviations from their long term monthly averages. Some studies have chosen rainfall alone, either by initial choice or because the other weather variables proved to be insignificant statistically. Fortunately, temperature and rainfall data are widely available.

Disaggregation of water use by customer class and/or area within the utility service area often leads to more accurate results. For example, a model of industrial water use might include variables such as the work force, value of product shipped, and some measures of technological change. A disaggregated analysis with an equation for each distinct group of customers may appear to be statistically weaker than a model with all users included, but will nevertheless provide more accurate forecasts of water use and the response to price and other independent variables.

5.9.2 FUNCTIONAL FORMS

If the model is estimated by multiple regression (ordinary least squares) in linear form, the estimated coefficient on each variable shows the slope of the relationship between that independent variable and quantity of water. In a model using monthly data, for example, suppose average Q = 5, average P = $1.95, and the coefficient on price (in dollars/1000 gallons) is -1.45. An increase in price of $0.20/1000 gallons would cause a cut in water use of 0.290 thousand gallons/month/customer (-1.45 x $0.2). The price elasticity is computed using average price and quantity over the entire study.

$$\text{Price Elasticity of Demand} = \frac{-1.45 \times 1.95}{5} = -0.57 \tag{5.15}$$

Therefore, for a 10 percent price increase in this example, Q would be expected to fall by 5.7% and revenue to rise by 4.3%.

There are various ways to specify and estimate non-linear models of demand. Non-linear models are often used for analysis because they can provide more accurate estimates. Some of the most useful non-linear models can be estimated using relatively simple data transformations. The widely used multiplicative model assumes that the effects of the independent variables multiply together instead of being additive as in the linear model. This model is estimated by taking the natural logarithms of each variable. The estimated model coefficients can then be interpreted directly as elasticities. For some situations, one of the semi-logarithmic models may be superior to either the linear or multiplicative (double logarithmic) models. Foster and Beattie (1979) used a semi-logarithmic model which made elasticity proportional to price in their path-breaking study. Table 5.2, shows the calculations of elasticity in each case.

Table 5.2 Calculation of Elasticities of Demand from Price Coefficients

Type	Equation	Elasticity
Linear	Q = a + b P	b (P/Q)
Semilogarithmic	Log Q = a + b P	b (P)
Doublelogarithmic	Log Q = a + b log P	b
Semilograthmic	Q = a + b log P	b/(Q)

Notes: a is the intercept term, b is the coefficient on price. Average values of P and Q are generally used in the calculation of price elasticity.

5.9.3 CROSS SECTION MODELS

Cross-sectional models for water demand are usually estimated using data for one year or one season, and include multiple water districts with differing prices. These models assume that differences in water use among the districts included in the study can be explained by the independent variables included in the model. Thus the choice of variables to include in the model is crucial to its success. Unless there are unusual elevation changes or coastal climate zones, cross-section studies of several water providers in a single metropolitan (or other small geographic) area do not require weather variables. The ideal economic model and the ultimately estimated multivariate (statistical) model may differ because of a lack of data for some variables. Cross section models generally include more variables than time-series models, in order to account for as many significant differences among the areas included as possible. If the model includes areas with significantly differing general price levels, the economic variables need to be adjusted for these differences. For example, if the study included San Francisco, California with its high living cost and Chicago, Illinois with its lower living cost, all prices and incomes would need to be adjusted to a common level. Such a model would also require climate and weather variables. If no price index is available for the included cities, an average wage rate index could be used.

Cross section models with a large number of independent variables often suffer from multicolinearity between two or more independent variables. Multicolinearity refers to the possibility that inclusion of one independent variable in the model will result in an incorrect estimate of the coefficient of one or more of the other independent variables. For example, if both personal income and home values are included as independent variables, they are generally highly correlated so that the estimated coefficients are not reliable. In this case, one of the two variables can be omitted. Multicollinearity is primarily a problem when the analyst is interested in the effect of one of the multicollinear variables. In many cases, the purpose of these variables is to control for differences among the areas included in the study and their individual results are of secondary interest. Since there are many econometric and multivariate books available, our book will not discuss the econometric details in estimating a statistical multivariate model. Given the large number of relatively inexpensive statistical and econometric software programs, many water providers should be able to create and maintain their own "in house" models at a relatively low cost.

5.9.4 USE OF SAMPLE SURVEY DATA

If the water provider can afford the cost, a sample survey can be used to gather data from customers for use in the economic and statistical models. For example, data could be gathered on the existence of low flow indoor water use devices, outdoor irrigation practices, swimming pools, number of occupants, age of the housing unit, and household income. The list of potential variables is very large and selection of the proper variables to include in the model is dependent upon the skill of the modeler, the budget for data gathering, and the time span in which the study must be concluded.

5.9.5 DEMAND MODELS FOR INCREASING BLOCK RATES

Many utilities have imposed increasing block rates that are often designed as conservation rates. These rate structures present additional challenges to the estimation of the price elasticity of demand for water. All price changes have both an income effect and a substitution effect. When the price of water rises, consumers experience an income effect because they are unable to buy as many goods and services as before because of higher water bills. However, when water bills are a small percent of consumer budgets, this effect will be quite small. Consumers also experience a substitution effect because the price of water relative to all other goods and services is now higher. When price is raised on a necessity good like water and income remains constant, the typical consumer will substitute a dirtier car, a less verdant lawn, and/or plumbing repair services for some prior water use. The consumer is also forced to reduce spending on other goods and services by enough to pay the now higher water bill.

With an increasing block rate structure, interior blocks with lower rates subsidize the purchase of water for consumers in the higher rate blocks. (High use customers pay less than their marginal price times total quantity purchased.) This produces an income effect of a price change that is not measured properly by the price variable as stated in the simple model of equation 5.7. Furthermore, a condition known as simultaneity bias is created by which the price influences the quantity purchased and the quantity purchased helps determine the price. The effect is to underestimate the price elasticity of demand (Martin et. al., 1981).

A variety of econometric approaches have been suggested for solving this problem, including doing nothing. One solution to the simultaneity bias problem is suggested by Agthe (et. al., 1986) in a study of residential water demand. A three equation simultaneous model is developed which includes both marginal price and a second price structure variable representing the income effect of the interior block subsidy and fixed fees. A set of dummy variables representing shifts in the rate schedule are included in the second and third equations. This method can be used for either average price (with two equations and one price variable) or marginal price (with three equations and two price variables).

Billings (1982) suggests an alternative method to remove simultaneity bias for studies utilizing marginal price. The procedure involves creating instrumental variables for both price and the interior block rate subsidy using water bills computed from each rate schedule and quantity.

Jones and Morris (1984) present an alternative method for creating the instrumental variables required to remove simultaneity bias. Their instrumental variable is the price of the average quantity consumed within their sample of households for each rate schedule. These prices, which can be either average or marginal, are then substituted for prices shown in the simple model.

An alternative method for solving the simultaneity problem for both marginal cost and average cost models using a discreet/continuous choice model is suggested by Hewitt and Haneman (1995). This approach involves first estimating which block of the rate schedule the consumer will be in and then estimating the response to price changes within that block.

5.10 Conclusion

Elasticities are useful for planning purposes. Particularly valuable is the insight that the absolute values of the price elasticity of demand and the price elasticity of revenue always sum to unity. For example, if the price elasticity of demand is -0.34, the corresponding price elasticity of revenue is +0.66. Thus a 10% rate increase is likely to result in water volume falling by 3.4% and revenue rising by 6.6%. Ignoring the existence of elasticity is equivalent to the assumption that its coefficient is zero so that price has no effect on quantity and revenue changes will be directly proportional to changes in water rates. This is a highly unlikely situation, and making this assumption leads to erroneous forecasts and sometimes severe embarrassment for the utility officials, as a rate increase fails to produce the anticipated revenue increases.

Elasticities are most useful when based on an economic/hydrologic model that is estimated with the best available data, although price elasticities which have been previously estimated can often be used when local data and estimates are unavailable. For an extended discussion of using elasticities to forecast, see Billings and Jones, 1996. Even if a local or regional elasticity study is not available, the analysis of the effects of rate changes should take elasticities into account.

References

Agthe, D.E., and R.B. Billings, 1997. "Equity and Conservation Pricing Policy for A Government-Run Water Utility," *Journal of Water Supply and Research Technology-Aqua*, 45 (5): 252-260.

Agthe, D.E., and R.B. Billings, 1987. "Equity, Price Elasticity, and Household Income Under Increasing Block Rates for Water," *The American Journal of Economics and Sociology*, 46 (3): 273-286.

Agthe, D.E., R.B. Billings, J.L. Dobra and K. Raffiee, 1986. "A Simultaneous Equation Model for Block Rates," *Water Resources Research*, 22 (1):1-4.

Babin, F.G., C.E. Willis and P.G. Allan, 1982. "Substitutes Between Water and Other Inputs," *American Journal of Agricultural Economics*, 64 (1): 149-151.

Billings, R.B., 1982. "Specification of Block Rate Variables in Demand Models," *Land Economics*, 58 (3): 73-84.

Billings, R.B. and D.E. Agthe, 1980. "Price Elasticities for Water: A Case of Increasing Block Rates," *Land Economics*, 56 (1): 73-84.

Billings, R. B. and C.V. Jones, 1996. *Forecasting Urban Water Demand*, AWWA, Denver, Colorado.

Clark, R.M., and H.C. Goddard, 1977. "Cost and Quality of Water Supply," *Journal of the American Water Works Association*," 69 (1): 3-15.

Gottlieb, M., 1963. "Urban Domestic Demand for Water: A Kansas Case Study," *Land Economics*, 39 (2): 204-210.

Griffin, R.C. and C. Chang, 1990. "Pretest Analysis of Water Demand in Thirty Communities," *Water Resources Research*, 26 (10): 2251-2255.

Griffin, A. and W.E. Martin, 1981. "Price Elasticities for Water: A Case of Increasing Block Rates: Comment," Land Economics, 57 (2): 266-273.

Grima, A.P., 1973. "The Impact of Policy Variables on Residential Water Demand and Related Investment Requirements," Water Resources Bulletin, 9 (4): 703-710.

Hewitt, J.A. and W.M. Hanemann, 1995. "A Discrete/ Continuous Choice Approach to Residential Water Demand under Block Rate Pricing." *Land Economics*, 71 (2): 173-192.

Houthakker, H.S. and L.D. Taylor, 1970. *Consumer Demand in American*, 2nd Edition, Harvard University Press, Cambridge, Massachusetts.

Howe, C.W, and F.P. Linaweaver Jr., 1967. "The Impact of Price on Residential Wader Demand and Its Relation to System Design and Price Structure," *Water Resources Research*, 3 (1): 13-32.

Jones, C.V., and J.R. Morris, 1984. "Instrumental Price Estimates and Residential Water Demand," *Water Resources Research*, 20, (2): 197-201.

Morgan, W.D., and P. Pelosi, 1980. "The Effects of Water Conservation Kits on Water Use," *Journal of the American Water Works Association*, 72 (3): 131-133.

Muncur, J.E.T., 1987. "Urban Water Pricing and Drought Management," *Water Resources Research*, 23 (3): 393-398.

Rensetti, S., 1988. "An Econometric Study of Industrial Water Demands in British Columbia," *Water Resources Research* 24 (10): 1569-1573.

Thornthwaite, C.W., and J.F. Mather, 1955. *The Water Balance, Publications in Climatology*,South Centerton, New Jersey, Drexel Institute of Technology, Laboratory of Climatology.

Witcomb, J.B., 1990. "Water Use Reductions from Retrofitting Indoor Water fixtures," *Water Resources Bulletin*, 26 (6): 921-926.

Weber, J.A., 1989. "Forecasting Demand and Measuring Price Elasticity," *Journal of the American Water Works Association*, 81 (5): 57-65.

Wong, S.T., 1972."A Model on Municipal Water Demand: A Case Study of Northeastern Illinois," *Land Economics*, 48 (2): 34-44.

Chapter 6

Water Transactions as an Urban Water Supply Strategy

Bonnie G. Colby
University of Arizona

6.1 Water Transactions in the Western U.S.: An Overview

In the early years of water market development in the western United States, urban growth was the primary motivation for water transactions. Municipalities have engineered various types of water transfers over the past several decades, as they have sought to supply new residential and commercial customers and to drought-proof their supplies. While water transfers continue to be important for urban growth, they have also become a valuable tool for environmental protection and restoration. Some environmental advocates embrace voluntary transfers as a means to encourage water conservation, stretch scarce regional water resources and to replace the hostility created by litigation and forced administrative reallocations with collaborative interactions. Native American tribal governments also have an interest in water transfers (De Coteau, 2000). In some cases, tribes lease their water to others, while in other instances they seek water for reservation needs. Recreation advocates urge transfers to provide water for whitewater rafting, sport fisheries and other recreational uses. Large-scale industrial users, such as power plants, mines, breweries and microchip plants buy, sell and lease water as their needs change. The agricultural sector is central to water transactions

D.E. Agthe et al. (eds.), Managing Urban Water Supply, 87–101.

because crop irrigation is the primary consumptive use of water in most of the west. Therefore, agriculture is the sector which others look to for water.[14].

6.2 Concerns About Urban Water Acquisitions

Agriculturally-linked rural communities and local governments (such as irrigation districts and counties) often respond with suspicion and alarm to cities' efforts to purchase water from farmers. They are concerned that the transfer of water from local agriculture will reduce business activity and property values and undermine the local economy. The example of Owens Valley persists over the decades (Reisner, 1986).

Water transfers away from agriculture have *both* positive and negative impacts. While the negative consequences for agricultural communities receive the most attention, transfers also generate negative third party effects.[15] Some of these include:

- Reduced profits for "backward-linked" businesses; those that sell inputs and services to farmers.
- Reduced profits for "forward-linked" businesses; those that purchase crops for their own use or for processing (feedlots, cotton gins) and now must turn to more distant supplies.
- Reduced profits for general businesses in the area that sell goods and services to households (assuming that water sellers exit the area, reducing the number of households).
- Reduced jobs in all of the businesses mentioned above.
- Reduced property values associated with a decline in businesses activity.
- Reduced tax revenues for state and local governments due to lower business sales and property values.

The potential for these negative impacts to occur depends upon whether revenues from water sales are reinvested in local farm operations or go to other businesses.[16] If farmers that sell water remain in the local area and spend most of their revenues there

[14] See Saliba and Bush, 1987; Anderson and Snyder, 1997; and The Water Strategist for examples of specific transactions.)

[15] Negative economic impacts include lost farm profits from reduced irrigated acreage. However, this is not a third party impact, since the seller must receive enough revenue from a voluntary transaction in order to willingly give up these farm profits.

[16] The magnitude of possible negative local economic impacts depends on whether the water reallocation:

- Reduces irrigated acreage. (Farmers may be able to continue to irrigate the same acreage by using water more efficiently.)
- Reduces farm revenues from crop sales. (Even if irrigated acreage declines, crop revenues may not decline due to increased yields on remaining lands, or a switch to higher value crops.)
- Decreases farm household income. (Even if crop revenues decline, voluntary transactions bring water payments to the farm household. Also, many farm households rely more heavily on off-farm income and on government payments than they do on crop sales.)
- Decreases spending in the local economy (Farmers selling and leasing water may invest in new irrigation technology and other farm improvements, or in off-farm activities in the local area. Moreover, the water may be used by other sectors in the local area, stimulating off-farm jobs ands economic activity.)

(on farm operations and other types of purchases), then the overall business activity and jobs in the region may increase. The *positive third party impacts* generated by the transfer of water from agriculture to urban and environmental uses may include:

- Increased economic activity in the sectors that acquire water. In the case of transfers to environmental purposes, there may be increased recreation expenditures in businesses that supply goods and services to hunters, anglers, bird watchers and other visitors who are drawn to the area by improved environmental conditions.
- Increased property values associated with new economic activities.
- Increased tax revenues collected by state and local governments on property values and sales.
- Increased recreational benefits to local residents that come with improved streams, wetlands and wildlife habitat.

Due to the variation in local economic consequences from water transfers out of agriculture, each case must be examined on its own merits. A number of researchers have examined actual transfers and modeled the effects of proposed transactions. Studies of these transfers find that local economic impacts are relatively small compared to the amount of irrigated land that is fallowed, even when the water is moved to a new use away from the area of origin (Nunn and Checchio, 1988; Howe, 1990; Coppock et al, 1994). The consistent findings of small impacts are due to several factors:

- Farmers fallow their lowest value crops and their least productive acreage;
- A portion of the water payments received by farmers generally is spent in the county from which the water is exported;
- Income from crop sales is a small portion of nearly every rural county's income in the west;
- Rural county households rely more on income from off-farm employment and government payments than from crop sales; and
- Water transfers can stimulate local off-farm jobs and income.

Even though the economic impacts due to removing water from agriculture are generally small, they may be concentrated in a few types of businesses and a few specific agriculturally-linked communities (Howe, 1990, 2000). Moreover, local concerns about impacts generate considerable political attention and can create obstacles to the plans of a city to buy water. Urban water managers do well to address such local concerns in order to avert the political and financial costs associated with opposing proposed water acquisitions.

Market transactions can be structured in ways that help to minimize negative third party effects and maintain agricultural activity in rural areas. One approach to preserving an agricultural base is to make transfers contingent on drought conditions. Farming will continue as usual in normal years and farmers can be adequately compensated in dry years. Another approach entails paying farmers to adopt water conservation practices and transferring only the conserved water. Also, cities can arrange to rotate the fallowed acreage (and the water lease payments) among many landowners (instead of permanently drying up a few farms) to maintain the baseline agricultural economy. Finally, purchasing only a portion of the water rights associated with a particular farm can leave adequate water rights to support gardens, horse pasture and a rural lifestyle, thus preserving property values. The legal ability to implement these strategies may vary with location, as state and local governments may impose restrictions on water transfers.

The parties most disconcerted by proposed transfers generally are not those who have water to sell. Farmers and irrigation districts that hold transferable water rights will look after their own interests when negotiating a water transfer, but there may be no forum for rural residents, local governments and affected local businesses. Some states have enacted legislation that allows local governments to review and approve proposed transfers of water out of their jurisdiction (National Research Council, 1992). In response to area-of-origin concerns, Oregon, Kansas, Nevada and Texas have each enacted additional procedural requirements for transfers and/or appropriations that would move water across basins or over long distances (Getches, 2000). States can protect a rural county tax base by requiring in lieu tax payments by public landowners when agricultural land is acquired by a municipality. States can also legislate the basis on which "water ranches" will be assessed (as a utility, or as agricultural land) for property taxation purposes. Some states attempt to regulate the nuisance effects of retired farmland by requiring the control of dust, weeds and insects. Arizona tumbleweed control legislation was passed in 1986 after the media published photos of a rancher unable to use his front door due to tumbleweeds piled up from retired farmland that had been bought by a city for water supply purposes (Saliba and Bush, 1987). Colorado requires the revegetation of farmland that is fallowed as part of a water acquisition (Getches, 2000).

In general, the economic impacts that changes in water use have on the area of origin can be addressed with specific legislation and through the structure of water transfers. However, beneath concerns over local economic impacts lie fundamental (and well-founded) concerns that water transfers signal a change in society's priorities and values for farms, cities, fish and wetlands. This resistance to social change stimulates rural objections to changes in water use, which are not necessarily remedied by compensation schemes and the restructuring of transfers.

6.3 Alternative Ways to Structure Water Acquisitions

Several decades of water transfers in the western United States have shown that the types of arrangements for transferring water from one use to another are limited only by the creativity of the parties involved. Possibly the most fundamental distinction is between voluntary and involuntary arrangements. Voluntary reallocations include purchasing and leasing water, incentive pricing, and technical and financial assistance to promote water conservation and release water for other uses. Compulsory mechanisms follow the three branches of government: court orders, administrative actions and legislative mandates. Examples include litigation and administrative actions to alter water diversions and change upstream dam releases as well as legislation that mandates improved water conservation practices. While there are important distinctions between voluntary and compulsory approaches, they complement each other in achieving changes in water use.

6.3.1 VOLUNTARY CHANGES IN USE

The western United States provides many examples of voluntary water transactions. The motivating force behind voluntary transfers is the perception that there may be economic gains to be made from transferring water to a new use in which it generates higher net returns. The different economic benefits per acre foot may be due to the variety of crops within an agricultural district, different economic values for water across the agricultural, municipal, industrial and environmental sectors, and various levels of willingness to bear the risk of water supply shortages. Necessity is still the mother of invention, and crises (long-term drought, prolonged litigation, contamination of a water source) are still the primary impetus for water transfers. One need only look at the changes made in California during the 1980s and 1990s to observe the innovations in water transfers that are inspired by several years of serious drought (National Research Council, 1992, Water Strategist, 1990-1998).

Three conditions must be satisfied for a buyer and seller to consummate a water transaction:

- The seller must receive a price offer that equals or exceeds the economic benefits sacrificed in transferring the water. An irrigator, for instance, must consider the net returns to water in irrigation, any decreases in the value of land due to reduced water availability and the expected appreciation in the value of the water right over time.
- The buyer must expect the benefits (including environmental benefits) from the water right purchase to exceed the costs.
- The buyer must view market acquisition of water as an economically attractive method of obtaining water, relative to other possibilities—such as litigation to change water allocation or new supply development.

While there are a few well-developed and active markets in the western U.S., water transactions generally are sporadic and complex. They deviate from markets for land and other real property in their "competitiveness." Water markets in the western U.S. typically are "thin," meaning there are only one or two major water buyers in a region, or there may be only a few potential sellers. It is not unusual for a water market to involve the one large city in an area as a buyer and a few farmers or a single irrigation district as potential sellers. In some areas, a water right acquisition may only occur every few years, while in active markets there may be a dozen transactions a month. In addition to negotiations between the buyer and seller, affected third parties may be included in the negotiations, and various forms of regulatory approval are required for many types of transfers (MacDonnell, 1990).

Transfers vary in duration from a permanent transfer to a seasonal lease to a "dry year only" option. There also are many different structures that can be used to establish price and to facilitate negotiations between potential sellers and those seeking to acquire water.

6.3.2 NEGOTIATED PURCHASES

Typically, prices are established and water transfers are negotiated on a case-by-case basis. Prices that emerge from bargaining between potential buyers/lessors and sellers/lessees will reflect the value of water to the parties involved and the overall demand and supply conditions in the basin. When multiple transactions involving similar types of water occur, a going price is established and this signals the value of water, encouraging water conservation and its transfer to higher-valued uses. With case-by-case negotiations, the parties can tailor the transfer to meet their specific needs with regard to the timing of the payments and the schedule to phase in new water use. A potential disadvantage of negotiated transactions is the costs incurred. There also is a tendency for negotiated prices to cluster around a norm established in past transactions, even when water values have changed (Colby, 1991). When this occurs, prices do not serve the desirable function of signaling changes in value.

6.3.3 AUCTIONS

Auctions have been used to allocate water supplies among potential water buyers, with recent examples in Victoria, Australia and in Texas (Water Strategist, 1997). Auctions have a number of advantages. They are public and transparent, which are desirable qualities when allocating water from public projects. If eligible to participate, environmental organizations and agencies can use auctions to acquire water for environmental restoration. These auctions rely on the forces of supply and demand to produce prices that reflect current market conditions and that respond to changing conditions. They do not require an administrator to set a price for water. Instead, the price emerges in the auction process and reflects the participants' knowledge of the basin and of the value they and others place on water in alternative uses. Variations in the attributes of water rights, such as seniority and location, are considered by bidders in the auction setting and reflected in the price.

The disadvantages of auctions can include a lack of familiarity with the process among potential water buyers and sellers, which creates the need to carefully design and describe the auction process so the rules and procedures are clear and collusion among participants to influence prices is thwarted. Features which need to be specified in designing an auction process include: the minimum and maximum amounts of water that can be offered for exchange, the minimum and maximum acceptable prices per unit of water, bidding procedures, necessary documentation to be provided by sellers of the rights, the qualifications of bidders (such as financial ability and location of intended water uses), and procedures for determining the winning bid and resolving disputes that may arise.

6.3.4 STANDING OFFERS

Standing offers are a widely-used mechanism for soliciting water supplies by public entities. For example, the City of Albuquerque maintained a standing offer of about $1,000 per acre foot for senior irrigation rights and gradually acquired supplies for urban growth in this manner over much of the 1980s. In the mid-1990s, the need to acquire water for urban growth became more pressing, so the standing offer price rose to $3,000 per acre-foot and in 2002 it was $3,800 per acre-foot (Water Strategist, March 2002). A standing offer involves publicizing a fixed offer price for water rights with specific characteristics that meet the buyer's needs. Standing offers have the advantage of simplicity and minimum transaction costs, since there is no negotiation over price; the potential sellers merely need to decide whether the posted price is adequate. The key disadvantage of standing offers lies in correctly setting the offer price and specifying which types of water rights will be accepted for acquisition. If the offer price is too low, this will become apparent as few takers will come forward and the price will need to be adjusted and re-advertised. If the price is set too high, more offers will be received than are needed and the purchaser will have to determine which water rights to acquire and which to reject, in addition to paying more than was necessary to acquire the quantity needed.

The complexities of a standing offer approach were illustrated in 1991-92 by the California Drought Emergency Water Bank. In 1991, the Bank offered $125 per acre-foot to willing agricultural lessors and quickly acquired 820,000 acre-feet. However, end users only wanted 655,000 acre feet at the price they had to pay, so the bank and its buyer of last resort (the State Water Project) were left with a substantial amount of unwanted and relatively expensive water. On the positive side, the bank rapidly acquired water for drought needs from a pool of previously unidentified sellers and coordinated delivery of water to end users, although it did not accomplish this at least cost. The following year, the bank offered farmers just $50 per acre foot and acquired 154,000 acre feet, all of which was passed on to end users (Coppock, et. al., 1994). A standing offer approach is inflexible over the short run, as it is politically and logistically burdensome to change the offer price many times over the course of a year. However, the price can be varied from one year to the next, reflecting new knowledge of water supply and demand.

6.3.5 WATER BANKS

Water banks can serve many important water management objectives, and can be valuable to cities seeking water. A water bank is an institutional arrangement for storing water to be used in the future and to facilitate trade and negotiations among those who contribute water to the bank and those who seek to use banked water. If water banks can standardize the units of water to be traded and the trading procedures, they reduce transaction costs because the participants need not independently locate trading partners and develop contracts, pricing and other terms of trade. A water bank can help match buyers and sellers based on their preferences. The bank's supply of water is an aggregation of the various types of water made available by sellers/lessors. This water is then packaged in ways that meet the timing and reliability objectives of buyers.

Water banks can use reservoir storage capacity, or can store water underground in aquifers. Reservoir storage entails negotiations with reservoir operators and others who use the reservoir. Potential conflicts involve determining "whose" water is spilled when flood release spills are necessary and "whose" water remains when stored water is low during drought. Underground storage also can be complex. Will water be recovered at the same site where it was recharged to the aquifer? Are there local impacts on nearby landowners from aquifer storage and recovery? Does the water migrate to another part of the aquifer? What are the risks of contamination?

The Metropolitan Water District of Southern California (MWD) has negotiated underground water banking arrangements with agricultural districts to allow some of the MWD's surface water supplies to be used by agricultural areas during years of plentiful supply. The groundwater not pumped by the irrigators during those years is then considered the MWD's stored water, available to them during years of low surface water supply (Water Strategist, 1996). Water banks also have operated Idaho's Snake River Basin for several years (MacDonnell et al., 1994).

The state of Arizona created the Arizona Water Bank in 1996 in order to promote direct use and groundwater storage of Colorado River water, imported into central Arizona via the Central Arizona Project. Surface water is banked in adequate water supply years, so that it can serve as a supply buffer in drought years. The bank seeks to drought-proof central Arizona cities and agriculture, and also allows California and Nevada to store excess water in Arizona aquifers. Water banked in Arizona by these states can be credited against excess surface water use in other years (Getches, 2000).

6.3.6 CONTINGENT TRANSFERS FOR DROUGHT PROTECTION

Dry year options and conditional lease-backs, negotiated in anticipation of drought, are two ways to ensure that water can be transferred quickly during drought. The difference between these two approaches is the ownership of the water. Under a *dry-year option,* ownership of the water right remains with the original water user. The new water user, a municipal water provider or a state agency, enters into an agreement with an irrigator allowing them to use water under specific conditions. For water users who need highly reliable supplies, this type of arrangement provides a back-up source of water for dry years.

Though they are promising, dry-year options can be difficult to negotiate with farmers who desire certainty when planning their farming operations. The following example illustrates this point. In 1987, the MWD attempted to negotiate a dry-year option with the Palo Verde Irrigation District (PVID). Under the proposed arrangement, the MWD offered an up front payment to Palo Verde farmers at the time they registered acreage for the dry year option, along with additional payments during the years the option would be exercised. The MWD expected to call that acreage into retirement about once every seven years in order to firm up municipal supplies. Farmers rejected the proposal for a number of reasons, including its effect on their ability to make long-range farming plans. In such arrangements, farmers face substantial uncertainty in planning their crop rotations, marketing strategies, equipment leases, and purchases of inputs.

A number of other issues need to be addressed when dry-year options are considered, such as defining the conditions under which the option will be exercised. Reservoir and stream flow levels can be specified as a basis for activating the option. Additionally, it is necessary to ensure that farmers be compensated for lost crop revenues when the option is exercised, disruption of farm planning and land use patterns, and for any production and marketing expenses incurred prior to being notified that land would be dried up for that season. The terms and timing of the notification are important issues to irrigators. In the early 1990s, the MWD and the PVID finally reached an agreement to help firm up MWD supplies during dry years (Water Strategist, 1992).

Under *conditional lease-backs,* land and water are purchased by the entity that wants to have long-term control of the water, most often a municipality or an industry, and are leased back to the farmer so that farming can continue except when the water is needed to replace drought shortfalls. The new water right holder could be a state agency, and the lease-back conditioned on the need for water to support instream flows during dry years. Conditional lease-backs are attractive to growing cities because they assure a supply of water that can be reserved either during droughts, or for water demand generated by new growth. Land and water acquisitions by the City of Mesa, Arizona provide an example of this type of arrangement (Saliba and Bush, 1987).

6.4 The Role of Litigation

In some regions, voluntary water acquisitions may not produce enough water to meet urban needs. Both litigation and voluntary mechanisms are commonly used to reallocate water in the western United States. Each approach has advantages and weaknesses that can complement one another. One of the key observations from researchers that analyze water conflicts is the interplay between compulsory and voluntary strategies to move water from one use to another (Kenney and Lord, 1999, Colby, 1998).

A market transaction and litigation may appear to be opposite tactics. The market approach accepts existing property rights without argument and fully compensates owners who sell or lease their land and water. Litigation forces changes in land and water management and may constrain property rights, or redefine rights altogether. However, the distinctions begin to blur on closer inspection. Like court rulings, market transactions can generate heated reactions. This occurs when the transaction negatively affects third parties, even though the buyer and seller have reached a mutually acceptable arrangement (National Research Council, 1992). "Voluntary" agreements often occur under the threat of a looming financial crisis, litigation or a regulatory change, and so may not be purely voluntary (National Research Council, 1992).

In many regions, both litigation and voluntary transactions are used to change patterns of water use and management.[17] Conflicts over California's San Joaquin River illustrate the way these approaches compliment each other. The San Joaquin River flowed unimpeded from the mountains to the Sacramento-San Joaquin Delta until the closing of Friant Dam in the mid-1940s. Since then, irrigation diversions in the San Joaquin Valley have created a dry stretch of nearly 20 miles (32 km) along the river's course (Water Education Foundation). Fifteen environmental groups, headed by the Natural Resources Defense Council, sued the Bureau of Reclamation as well as the irrigation districts for violations of the state Fish and Game Code and the resulting damage to the riparian ecosystem and fish populations (Natural Resources Defense Council v. Houston, 1998). The 9th U.S. Circuit Court of Appeals ruled in favor of the plaintiffs on major issues and voided agricultural water district contracts with the U.S. Bureau of Reclamation for Friant Dam water. In response, the parties agreed to undertake pilot restoration projects. The restoration project has changed water release patterns to benefit the riverine habitat but has not decreased the net supply of water for the FWUA, due to exchanges negotiated with water users in other parts of the San Joaquin Valley. Several million dollars have been made available to purchase replacement water for irrigators (Water Education Foundation, 1999). The resolution of the dispute between environmental advocates and irrigators takes the pressure off nearby cities which otherwise might have faced pressure to reduce their water consumption in order to provide water for environmental needs.

Without the threat of litigation, voluntary agreements in which irrigators alter their customary uses of water to accommodate other water needs are difficult to achieve.[18] Compulsory mechanisms provide the stimulus for negotiated agreements and voluntary transactions. Consequently, city water managers should be alert to the opportunities and perils posed by litigation over water in their region.

6.5 Policy Changes to Facilitate Water Acquisitions

The western states and the federal government have a long legacy of policymaking regarding water transfers.[19] Western state policies have vacillated between opening up water markets and encouraging transactions on the one hand, and imposing new restrictions and procedural requirements on changes in water use on the other hand. This is not surprising as state agencies and elected officials struggle to balance the water desires of the new west (lushly landscaped and rapidly growing cities, abundant water-based recreation and restoration of species and their habitats) with agricultural and rural consternation over the fading of the old west and its customary water use patterns. Getches (2000) observes that most state policy changes of the 1990s served to restrict

[17] See, for instance, the Pyramid Lake case as described in National Research Council, 1992, pp. 119-136.

[18] See, for instance, the cases documented in d'Estrée and Colby, 2000; Kenney and Lord, 1999; Checchio and Colby, 1992 and National Research Council, 1992.

[19] For an overview of policies in the 1970s, 1980s and early 1990s, see Driver, 1986; Mac Donnell, 1990; National Research Council, 1992; Reisner and Bates, 1990; Wahl, 1989 and Willey and Graff, 1994.

water transfers and few changes were promulgated to open up and broaden water market activity.

While water has been moved to new municipal, industrial, recreational and environmental uses in a dizzying array of arrangements over the past twenty years, there is still much that could be done. The following items are suggested policy reforms to facilitate beneficial transfers of water.

- Continue to refine the conditions under which western states define water as "conserved" and available for transfer to other uses. Clear rules would allow cities to pay farmers to implement water conservation, in return for use of the water conserved.
- Integrate water quality concerns into the policies that govern changes in water use – both as a valid reason to transfer water, and as criteria for evaluating proposed transfers. This can assist cities struggling to comply with surface water discharge standards for their wastewater treatment facilities, as well as those seeking to protect drinking water sources.
- Modernize federal and state policies to effectively recognize the interconnections between surface water and groundwater, and to account for these linkages in evaluating proposed water transfers. This will assist cities to establish a reliable water portfolio composed of surface water and groundwater resources.
- Establish basin-wide, multi-jurisdictional water transfer mechanisms as a way to respond to challenges such as drought, species restoration and improving water quality. Tribal, state and local governments may be able to share information and resources for improved water management. Cities can provide the leadership for cooperation on regional water issues (Weatherford, 2000).

6.6 Criteria for Desirable Urban Water Transactions

City water managers need criteria in order to identify desirable water transactions. The criteria below are derived from a larger framework for evaluating different strategies to resolve environmental disputes (D'Estree and Colby, 2000). Only a subset of those criteria is discussed here, as adapted to evaluate urban water transfers.

6.6.1 POSITIVE NET BENEFITS

This criterion asks whether the water reallocation creates positive net benefits that would not have been available otherwise[20]. The city may gain net benefits from additional growth supported by new water and from reduced exposure to drought risk. Net benefits also may arise from avoiding the costs of prolonged litigation, improved water quality, and from better regional sharing of water information and technology.

[20] This concept of weighing benefits and costs is central to the "mutual gains" negotiation framework described in *Getting to Yes* (Fisher, Ury, and Patton, 1991) and applied to environmental disputes in Susskind et al. (2001). It is sometimes called "creating value" or converting zero sum negotiations to positive sum negotiations.

Voluntary water acquisitions nearly always provide improvements for those who sign on, compared to their Best Alternative to a Negotiated Agreement (BATNA). Otherwise, the buyer and seller (or lessor and lessee) would decline to bind themselves to the water transfer agreement (Fisher, Ury, and Patton, 1991).

A thorough evaluation of a proposed transfer should consider not only the benefits and costs to the direct participants (the litigants and negotiants), but also the benefits and costs to public agencies, taxpayers and more dispersed interests who may not be at the table, such as water utility ratepayers, recreationists, and property owners in the affected region.

6.6.2 FAIR DISTRIBUTION OF COSTS AMONG PARTIES

"Fairness" appears consistently on lists of desirable characteristics for water transfers. Ideally, those undertaking a water transfer will first identify the most cost-effective plan to accomplish their water reallocation objectives (ie., provide 10,000 AF for fish recovery or improve city drought supplies by twenty percent). Then, they can use cost-sharing principles and compensation packages to even out discrepancies between the parties who gain from the transfer and those who face losses. There are many possible cost-sharing principles. One principle, for instance, involves sharing costs in proportion to the benefits received from the change in water use (beneficiaries pay). Another possible principle is to share costs in proportion to past damages to the ecosystem (biggest water users pay), or bear costs proportional to one's financial assets (deep pockets pay). There may be opportunities to assess dispersed interests who may not be direct parties to a water transfer agreement for some costs – recreationists, for instance, who will benefit from river restoration and could pay higher access fees.

6.6.3 PARADIGM SHIFT

From an urban water manager's perspective, paradigm shifts are desirable when they change outdated ways of managing water that keep water locked in agricultural uses. One of the essential elements in a paradigm-shifting event is that it changes the bargaining power among stakeholders. For instance, a court ruling that favors instream water use puts environmental advocates in a stronger bargaining position for all future negotiations. Court-rulings in favor of dedicating water to environmental uses not only address the litigated dispute, but also set important precedents. Voluntary water transactions can be paradigm shifters when the transaction is the first of its kind in a region: the first purchase and transfer of a senior water right to maintain river flows through a city park, for instance. Innovative transactions can pave the way for more widespread use of the market to accomplish changes in water use by forcing policy makers to clarify how traditional water right transfer processes will be applied to a new purpose, such as river restoration or drought-proofing a city.

6.6.4 ENHANCED SOCIAL CAPITAL

Social capital refers to improved cooperation and problem-solving capacity and the ability to fulfill mutual obligations. Social capital affects economic and political transactions by altering the terms of agreements reached, and the costs of achieving those agreements and monitoring and enforcing their implementation. An enhancement in social capital can reduce transaction costs such as attorneys' fees and improve the productivity of other inputs such as time spent in negotiations (North, 1990, Wilson, 2000).

Trust and reputation as elements of social capital are important factors in bargaining situations (Casson and Wilson, 2000). Enhanced trust can lead to easier and lower cost problem-solving. The stakeholders engaged in managing watersheds often must address multiple resource problems over a period of years. For instance, the dilemma this year may be over providing water for new growth, but in the next few years the same stakeholders may confront a drought or a drinking water quality problem. Consequently, their ability to build trust and work together effectively can be an important asset. Negotiated water transactions provide clear advantages over litigation because stakeholders become engaged in identifying strategies to accomplish changes in water use and building consensus for a particular approach. The bargaining process gives the stakeholders experience working together which can make it easier to solve the next water problem that faces the region. In contrast, litigation encourages an adversarial approach among the parties rather than a problem-solving stance.

6.7 Summary and Conclusions

Water acquisitions are an essential tool for managing the quantity and reliability of urban water supplies. However, water transactions typically involve negotiations among multiple interests and often are resisted by rural communities. Recent examples from the western U.S. illustrate innovative ways to structure water acquisitions so as to reduce third party impacts, while still accomplishing urban water management objectives. Litigation provides an important impetus for voluntary water transfers. State and federal policy initiatives could enhance cities' abilities to implement transactions. Finally, city water managers need to be able to distinguish desirable transactions from among the many possible acquisitions they consider. Desirable transactions are those which produce positive net benefits, distribute costs fairly among affected parties, develop new paradigms for water allocation and management, and enhance regional cooperation and problem-solving capacities.

References

Anderson, T.L, and P. Snyder, 1997. *Water Markets: Priming the Invisible Pump*, Washington D.C.: Cato Institute: 47-66.

Checchio, E. and B.G. Colby, 1992. *Indian Water R*ights, University of Arizona.

Colby, B.G.,1998. Negotiated Transactions as a Conflict Resolution Mechanism:72-94 in M. Rosegrant, A. Dinar, and W.K. Easter eds., *Markets for Water - Potential and Performance*, Kluwer Academic Publishers.

Colby, B.G., 1991. "Recent Trends in Southwestern Water Values," *Appraisal Journal*, 59:488–500.

Colby, B.G., 1990 "Transaction Costs and Efficiency in Western Water Allocation," *American Journal of Agricultural Economics* 72: 1184-1192.

Congressional Budget Office, Natural Resources and Commerce Division, 1997. *Water Use Conflicts in the West: Implications of Reforming the Bureau of Reclamation's Water Supply Policies*, Chapter 4, US Government Printing Office, Washington, DC.

Coppock, Ray et al., 1994. "California Water Transfers: The 1991 Drought Water Bank", in *Sharing Scarcity: Gains and Losses in California Water Marketing*, H. Carter et al, eds, University of California.

DeCoteau, J., 2000. "The Effects of Development on Indian Rights: Obstacles and Disincentives to Development of Indian Water Rights," paper presented at Natural Resources Law Center conference, University of Colorado School of Law, June.

D'Estrée, T. and B. Colby, 2000. "Guidebook for Analyzing Success in Environmental Conflict Resolution," Working Paper, Institute for Conflict Analysis and Resolution, George Mason University, Fairfax Virginia.

Driver, B., 1986. Western Water: *Tuning the System*, Report to the Western Governor's Association from the Water Efficiency Task Force 25.

Fisher, R., W. Ury and W. Patton, 1991. *Getting To Yes*, Penguin Books, New York.

Getches, David H., 2000. "The Metamorphosis of Western Water Policy: Have Federal Laws and Local Decisions Eclipsed the States' Role?" *Stanford Environmental Law Journal*, 20(1, January): 3-72.

Howe, Charles, et al., 1990. "The Economic Impact of Water Transfers in the Area of Origin: A Case Study of the Arkansas River Valley in Colorado", *Amer. Journ. Agricultural Economics*, 72 (5, December): 1200-1209.

Howe, Charles W., 2000. "Protecting Public Values in a Water Market Setting: Water Markets to Increase Efficiency and Equity," *U. Denver. Water L. Review*, 357: 368-72.

Kenney, D. and W. Lord, 1999. *Analysis of Institutional Innovation in the Natural Resources and Environmental Realm: The Emergence of Alternative Problem-Solving Strategies in the American West*, Natural Resources Law Center Research Report, No. 12, University of Colorado School of Law, Boulder: 26-52.

MacDonnell, L., 1990. *The Water Transfer Process as a Management Option for Meeting Changing Water Demands*, report to the U.S. Geological Survey, Natural Resources Law Center, University of Colorado Law School.

MacDonnell, L. et al, 1994. *Water Banks in the West*, Natural Resources Law Center, University of Colorado Law School.

MacDonnell, L. and T. A. Rice, 1994. "Moving Agricultural Water to Cities: The Search for Smarter Approaches," *N.W. J. ENVTL. L. & POL'Y*, 27.

National Research Council, 1992. *Water Transfer In the West*, National Academy Press: Washington D.C.

Natural Resources Defense Council v. Houston, 146F.3d. 1118,9th Cir., 1998.

North, D., 1990. "Institutions and Transaction Cost Theory of Exchange," in *Perspectives on Political Economy*. J. Alt and K. Shepsle, eds, Cambridge England: Cambridge University Press, 1990: 182-194.

Nunn, S.C. and E. Checchio, 1988. "Water Transfers in Arizona: Assessing the Impacts on Areas of Origin", *American Water Resoures Assoc. Journal*: 547-559.

Reisner, M. 1986. *Cadillac Desert*, Chapter 2, Viking-Penguin, New York.

Reisner, M. and S. Bates, 1990. *Overtapped Oasis: Reform Or Revolution For Western Water*, Island Press, Washington, D.C.

Saliba, B.G. and D.B. Bush, 1987. *Water Markets in Theory and Practice*, Westview Press, Boulder.

Susskind, L., Levy, P. and Thomas-Larmer, J., 2000. *Negotiating Environmental Agreements*,: Island Press, Washington, D.C.

Wahl, R. W., 1989. *Markets For Federal Water: Subsidies, Property Rights, And The Bureau Of Reclamation.*

Water Education Foundation, Sacramento, 1999. "Rewatering the San Joaquin River," *Western Water* September/October: 3.

Water Strategist, 1990-2002. Rodney Smith, editor, published by Strat Econ, Claremont, California.

Weatherford, G., 2000. "Regionalizing Water Management: The Evolving Hydrocommons", paper presented at the Natural Resources Law Center Conference, University of Colorado School of Law, June.

Wilson, P., 2000. "Social Capital, Trust and the Agribusiness of Economics," *Journal of Agricultural and Resource Economics* 25: 1-13.

Willey, Z and Tom Graff, 1988. "Federal Water Policy in the United States—An Agenda for Economic and Environmental Reform," *Colum. J. Envtl. L.* 13: 325.

Chapter 7

Market Structure and Price Regulation

Donald E. Agthe
Economic Consultant

7.1 Introduction

Since most urban water is sold in a monopoly or single seller market, it is not usually thought of as a candidate for market analysis. This is also true of water sources that are frequently allocated on the basis of inter-government agreement and/or water compact. For example, the Colorado River Compact allocates water between the upper and lower basin states and in the lower basin among the states of Nevada, Arizona, and California. Bonnie Colby discusses markets involving water sources in Chapter 6. This chapter emphasizes consumer markets.

Urban water providers usually face several distinct consumer markets and the mix of sales across these markets is important to the revenue potential and equity issues facing the urban water provider. These markets are usually referred to as customer classes. Furthermore, these customers usually do not engage in resale to other customers because it is uneconomic or illegal. Often the urban water utility monopoly is protected further by prohibitions on private wells within its market area, thus preventing the loss of customers to private wells. For a utility using groundwater from within its market area, the prohibition of private wells also eliminates competition in using the aquifer. In many urban areas, governments grant what amounts to an exclusive monopoly franchise in return for regulation and limitations of the profits which may be earned.

Standard economic theory suggests that an unregulated monopolistic producer will maximize profits by producing the quantity where the marginal cost of production is equal to marginal revenue in each of its several markets. If these markets are separated

D.E. Agthe et al. (eds.), Managing Urban Water Supply, 103–113.

(resale between consumers is impossible or illegal), the unregulated producer will charge the highest price to the least price elastic market and the lowest price to the most price elastic market to maximize profit.

This standard economic theory of profit maximization by an unregulated monopoly does not hold in most urban water markets because the water providers are government owned or government regulated private firms. This system of ownership and regulation arose because of the cost structure of water providers and the large positive public health externalities associated with water supply. Consequently, the local water provider's objectives are likely to be covering costs, making the allowed rate of return on invested capital, equity and conservation of water rather than profit maximization.

7.2 Advertising by Water Providers

Being monopolies, water providers are unlikely to use advertising to expand markets or increase profits because they are nonprofit or restricted profit enterprises and because water is becoming an increasingly scarce resource that needs conservation rather than expansion of uses and market. However, advertising can be used to promote special services, obtain approval of bonds to finance expansion and updating of the water system, improve the utility's public image, and to promote water conservation. Advertising to make people aware of conservation practices such as appropriate times to water, checking for leaks, installation of water saving landscapes, and general conservation awareness is increasingly common in many areas of the United States. Advertising can also promote special services such as free analysis of water use for residential and industrial customers, reduced water prices for water saving landscape installation, and free distribution of low-flow showerheads.

7.3 Customer Classes and Rate Design

Designing rate structures by customer class characteristics can help a utility meet its revenue and sales volume goals. The utility benefits by responding to differences in demand functions, with their corresponding marginal revenues and price elasticities among customer classes. For example, conservation rate structures designed for individual households are inapplicable for common metered multiple unit dwellings. Similarly, if water conservation is important and recycling of treated effluent is technically and economically feasible, the water provider may wish to price potable water relatively higher in the industrial, landscaping and golf course markets. The relatively higher price would encourage these customers to switch much of their water purchases from potable water to treated effluent.

Economic quantity of water and physical quantity of water are different concepts. The economic quantity can exceed the physical quantity because water can be reused. The economic quantity of water depends upon its use and the ability to recycle the water after use. For example, water could be used by residential consumers and later reused in the form of treated effluent on golf courses. Prioritizing the use of water according to customers' quality requirements increases the economic quantity of water. Market analysis of customer classes can be used to identify customers who need high quality potable water and those with needs that can be satisfied by lower quality water. This analysis is required to develop rate structures within each customer class and across customer classes that encourage the use of recycled water. Distribution costs tend to be high for recycled water because the recycled water cannot share the distribution system with potable water (a dual water system is required). The market and customer class analysis should also identify geographic clusters of customers who could be served by the secondary supply system.

Customer classes allow the water provider to gain information that can be used to target conservation policies or special services. For example, knowing whether people own or rent can help in designing an indoor conservation policy. Common metered apartment renters have little incentive to save water. Apartment renters are unlikely to invest in water saving capital such as low flow toilets or faucets because they typically do not directly pay for water. Even when the apartments are individually metered, the renters seldom live in an apartment long enough to recover these capital investments through water bill savings. Furthermore, renters have little control over outdoor water use.

Owners and managers of apartments should be targeted in water conservation programs aimed at common metered complexes for they are the ones to invest in indoor and outdoor water saving devices and practices (Agthe and Billings, 2002). Examples of these practices would be water saving landscaping, low-flow toilets and faucets, and covers for swimming pools. Cooperative and condominium owners have some voice in their homeowners associations to support water saving investments. They also have tenure of ownership that is an incentive to invest in their own unit.

7.4 Derived Demand: Commercial and Industrial Markets

While water is popularly thought of as being a final consumer good, in most communities over 80 percent of water is used in production of water related goods and services. In this broad sense, production includes household water uses such as "producing" clean clothes and grass lawns. The demand for water as an input to production is called derived demand. This is because the demand is derived from the place of water in the production of the final product and the demand for the final product. Technological innovation in production may either increase or decrease the associated derived demand for water. There is no reason to believe that technological progress in water using industries will necessarily help alleviate water supply problems. A change in the demand for the final product will directly affect the demand for water use in producing the good. A change in the price of a substitute input will also change the demand for water. Less water will be demanded for clothes washing, for example, if the price of water-conserving front loading machines falls relative to the price of higher water use top loading washers.

Car washes, clothes washing, landscaping, swimming pools, vegetable gardens, and decorative pools and fountains are examples where water is an input to another consumer good in the context of residential demand for water. In these cases, clean cars, clean clothes, swimming, fresh produce and attractive architecture are the corresponding products.

Almost all water used in golf courses, industrial processes and public recreational facilities represents derived demand. Water can be used for landscaping surrounding factories, schools, and industrial plants in the same manner as that for private homes. Public and private recreational facilities such as golf courses, sports fields, and swimming pools, use water as an input to produce a final good (i.e.: playing golf, football games or recreational swimming). Water is useful for industrial cooling, as a physical input to a final product such as soft drinks, or as part of the production process and product as in the chemical industry.

DeRooy (1974) developed the following economic model shown in equation 7.1 to explain industrial water demand for cooling, processing and steam generation. This model yielded price elasticities of demand for water of -0.89 for cooling, -0.75 for processing and -0.74 for steam generation. DeRooy found that the demand for water for sanitation was linearly related to the number of employees; a condition similar to that for commercial establishments and residential customers.

$$QW = f(QO, PI_1, \ldots, PI_n, PW) \qquad (7.1)$$

Where: QW = Quantity of Water
QO = Quantity of Output (market)
PI = Price of Other Inputs
PW = Price of Water

Babin (et al., 1982) examined derived demand for entire industries. In their model, the quantity of water consumed by firms depended upon capital expenditure, water price, production employment, and administrative employment. They found price elasticities of demand significant at the .95 level for several industries, including -0.66 for paper and allied products, -0.41 for metal fabrication, and -0.38 for stone/glass/clay. They also present cross price elasticities of substitution between pairs of inputs that measure response to price changes of each input. Renzetti (1988) calculated derived demand price elasticities for groups of heavy and light industries. Since zoning in many cities separates heavy and light industry, it is reasonable for customer classes to exist for each type of industry as their water demands are likely to be different. Renzetti found price elasticities of demand of -0.25 for heavy industry and -0.54 for light industry. In calculating these elasticities, Renzetti's independent variables were the price of water at intake, the cost of treating the water prior to use, the cost of re-circulation, the cost of treatment after use, and the output level. One problem encountered was that many of the water providers utilized block rates. To solve the problem of simultaneity bias, Renzetti created a series of instrumental variable intake prices by regressing the actual prices paid on rate schedule characteristics, a technique suggested by Jones and Morris (1984).

Industrial users can be targeted to receive special services to meet their individual needs and to work with the water provider in conserving water through product or process changes. For example, it is often possible for the water provider to demonstrate to the industrial firm that recycling water used for cooling purposes can save the community water and the industrial firm money. Experts might be brought in to show how water use can be reduced in other parts of the production process in such a way that the industrial firm's cost of water can be reduced. Golf courses can be targeted to substitute recycled effluent for potable water, change their hours of watering, or change their landscapes to less water using vegetation. Commercial users like restaurants can be targeted for policies to conserve drinking water. Many hotels have adopted policies to encourage guests to use the same bedding and towels for more than one day to save water used for laundering. Gas stations and apartment owners can be encouraged to sweep their sidewalks with a broom instead of a hose! Recreational and government users such as schools, public buildings, and parks can be given rate structures and use targets based on their water requirements. Separate pricing structures can be arranged if the process or product uses considerable water but is also important to the income and employment situation in the local economy. In general, customer classes allow the water provider to better know its customer's characteristics so that it can target water conservation programs, special rate structures, and/or special services.

The concept of water managers working with water using industries may be realistic. Smith (1998) points out that his experience with environmental problems in New York State led him to believe that many conservation problems caused by firms were due to attitude and ignorance. He found that materials, product and process changes often easily solved the problem and in most cases the firm was better off financially after making them. The same may be true for water in industrial processes. An astute water manager can point out what is necessary and educate the business firms' owners on how to conserve water in their product, materials and process management.

7.5 Marketing Channels

A marketing channel depicts the process of producing and marketing a product from its source to its final consumption. Most products have marketing channels associated with them. A typical urban marketing channel for water is depicted in Figure 7.1, including the water source, conveyance to market, treatment, delivery, and a final consumer.

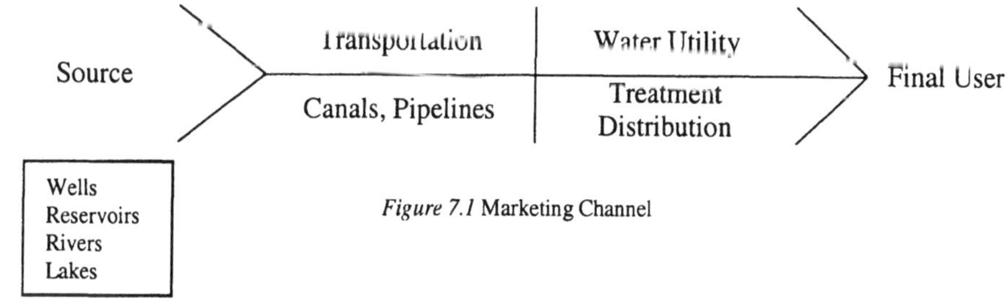

Figure 7.1 Marketing Channel

The simplest channel is that of a home in rural area that has a well as its source so the source and the customer exist in the same location. On the other hand, southern California cities have local wells and bring in additional water from the Colorado River and Northern California in extensive aqueduct systems; a very long marketing channel in number of steps and in mileage. While most marketing channels for water are still relatively short, lower water tables resulting from overdraft of aquifers will force many urban and suburban areas to develop distant sources. Thus, the marketing channel is likely to become a more important consideration of urban water providers over time.

Examining the marketing channel for water is useful to the water provider because it shows steps in the water provision process where potential markets might exist. It also shows where it might be possible to engage in horizontal or vertical integration to reduce the cost of provision of water. Vertical integration refers to a single entity carrying out multiple steps in the marketing channel. Many utilities are fully integrated vertically, although others buy water from regional wholesale suppliers. Horizontal integration refers to the practice of two or more water providers combining to provide a particular stage of the production process.

The potential for horizontal integration exists at most stages of the marketing channel. At the source stage, several water providers might combine to build a reservoir to provide surface water. At the transmission stage of the marketing channel, a number of water providers could share the cost of a canal or pipeline from a distant water source. A higher level of government or a private pipeline company could also provide this service. Several water providers may jointly build and operate a regional water treatment facility. Some water providers share administrative costs with other providers or outsource these tasks to a company that specializes in payrolls, employment and financial services. A government owned water provider could also share these tasks with other administrative units unrelated to water provision. It is not uncommon in the United States for small local water providers to be owned by a corporate group based far from the actual water markets faced by the providers. The economies of scale in engineering and management services make such companies attractive to entrepreneurs.

Markets for water can exist at several points in the marketing channel. The most likely location is the point of final delivery to the consumers, differentiated by customer class. Markets for water also potentially exist at water sources and at the end of the transmission pipeline. It may be possible for urban areas to bid for the water at each point or to sell water to other water providers at each of these points. Markets also are possible for services required in the production of water. If the water utility is too small to achieve meaningful economies of scale, services such as management, treatment, and engineering design are often bid out to private consultants or other government agencies to reduce their cost. On the other hand, a larger water utility could achieve economies of scale in production of the service by selling these services to other water providers.

7.6 Cost Structure and Provider Mergers

Water providers are frequently referred to as natural monopolies because of their heavy investment in fixed capital. This means that they have declining cost structures for their long run marginal and average costs. However, it is not uncommon for very large cities to have an increasing long run average and marginal cost structure as more raw water, labor, and other inputs are purchased.

In many urban areas individual wells are drying up as groundwater overdraft occurs, forcing homeowners and others to seek service from large urban water providers. Metropolitan areas such as Tampa, Florida; Phoenix, Arizona; and Los Angeles, California are becoming more dependent upon expensive distant water sources to supplement dwindling local supplies. The use of expensive and distant water sources often force the merger of small local water providers or the creation of a regional water authority such as the Metropolitan Water District in Los Angeles and the Central Arizona Water Conservancy District in Arizona. Most individual urban providers lack the financial resources to obtain and transport water from distant sources to the metropolitan area.

7.7 Rate and Service Regulation

Water providers are usually geographic monopolies either operated by the government or by regulated private companies. They are usually natural monopolies because of their cost structure. Multiple providers offering service to the same customers would mean duplication of the distribution system and an increase in total cost of water.

The regulated private providers are usually overseen by a government agency with respect to maintenance of minimum quality of service and reasonable rate structures. The regulated companies are allowed a "normal profit" (to satisfy their private investors) in exchange for their monopoly market. These restrictions are imposed to prevent the water provider from earning a monopoly profit. In the United States over one half of the water providers are privately owned but they only supply about 16 per cent of the population (see Chapter 11). In contrast, the United Kingdom opted for 100% privatization of the drinking water industry in 1989.

Both government run and regulated private monopolies have the potential to offer poor service to their customers. A protected market and lack of profit incentives often lead to inefficiencies such as overpaid administrators, over staffed administrations, a slow rate of technical innovation, and poor customer service. The slow rate of technical innovation may stem from using existing technology as a given, a fear of the higher cost of new technology, a fear of new technology becoming obsolete before it exceeds its useful economic life, fear of technical complexity, and the lack of effective incentives for innovation. The lack of incentives for technologic innovation is most harmful when improved technology protecting against water borne diseases is installed slowly or perhaps not at all. Furthermore, even in non-health related activities such as billing, metering, or construction there is little incentive to adopt the most economically efficient methods and technologies. Gray (2003) provides many examples of cost-saving water plant design innovations which utilities are loath to adopt because these innovations are considered unconventional. Cost plus engineering and construction contracts may also provide incentives toward more expensive technologies.

Governmental water utilities in North America generally follow a tradition of being nonprofit providers. This differs from local government owned electric utilities that often charge rates that include a profit that is used to offset property taxes. The nonprofit and limited profit models for water provision frequently prevent the water provider from charging conservation rates or even rates that represent economic efficiency.

7.8 Rate Setting: Should Average or Marginal Cost Determine Price?

Economic efficiency requires that the commodity price of water be set equal to its marginal cost of production.[21] However, regulators attempting to meet this requirement face a possible dilemma as shown in figures 7.2 and 7.3. These figures show a typical "U" shaped long run average cost curve (ATC), its corresponding marginal cost curve (MC), which always intersects ATC at its lowest point, and the downward sloping (negative sloped) demand curve (D) for a monopoly. Average total cost in this case includes water development costs and a normal rate of return as well as the capital and operating costs of the water provider.

In Figure 7.2, setting the commodity price where the demand curve intersects the marginal cost curve results in selling a quantity of 9 units at price P, which is below average cost at that quantity, resulting in a loss. Figure 7.3 depicts the opposite possibility: setting the price where demand intersects the marginal cost curve results in selling 23 units and a profit, since the price at 23 units is above the average cost of production.

Given that the quantities where demand intersects marginal cost are economically efficient, following the traditional policy of setting price equal to average cost is inefficient. Too little water is sold using average cost under the conditions depicted in Figure 7.2 and too much water is sold using average cost under the conditions shown in Figure 7.3. In the extremely rare case of the demand curve cutting both ATC and MC at their intersection the regulator would have no dilemma and all costs would be covered with only a normal rate of return being earned.

Traditionally, regulators and administrators have required that prices be set so that total revenue is approximately equal to total cost, which includes a normal rate of return for a privately owned water provider. In many rate schedules, periodic service fees are set to cover costs that do not vary with the quantity of water provided while the commodity charge for water is set equal to its average cost of provision. Average cost is used in setting price because it is easier to understand and compute than marginal cost. However, in Figure 7.3, the efficient quantity of water is smaller than the amount that is

---

[21] Water rates for blended water can represent an average of the cost of obtaining water from a number or sources. However, economic efficiency requires that price be set equal to the marginal cost of the most expensive source.

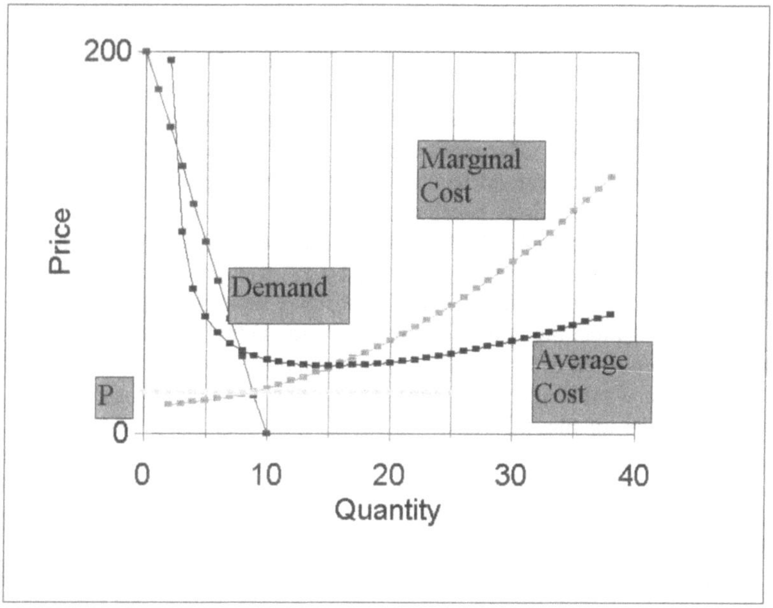

Figure 7.2 Regulated Monopoly Price < Average Cost; for Price Set Equal to Marginal Cost

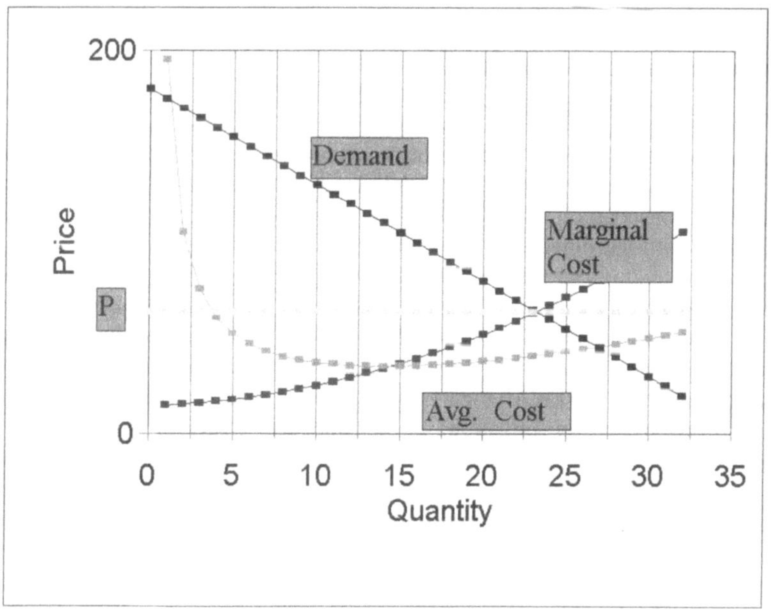

Figure 7.3 Regulated Monopoly Price > Average Cost; for Price Set Equal to Marginal Cost

chosen where average total cost intersects the demand curve. In this case the consumer receives a lower price and a larger quantity of water under average cost pricing (price equals average total cost) than at the economically efficient option (price equals marginal cost). Setting price equal to long run marginal cost in this case generates surplus revenue and profit. The profit could be used to reduce the periodic service fee or even result in a periodic refund to customers. Annual refunds might result in lower water use than rate reductions because of consumer cash flow limitations or because refunds would be thought of as "found money" rather than as a reduction in the cost of water. The surplus could also be taxed away by government. Some government owned utilities could use surplus utility revenue to subsidize their general budget.

Figure 7.2 depicts the opposite case where the actual quantity under average cost pricing falls short of the economically efficient quantity. In this case, setting price equal to marginal cost would not cover total cost, so that the periodic service fee could be increased to make up the difference. If a higher service fee is not allowed, water is unlikely to be provided without a subsidy. A private supplier would require an explicit subsidy, while a government agency could quietly cover costs such as capital investments from other revenue sources. However, water use would be efficient in terms of price and quantity.

Timmins (2002) investigated the losses attributable to improper price setting for water by regulatory and municipal agencies in the Western United States and found them to be substantial. He points out that inefficient exploitation of the aquifer results from under-pricing and that society incurs great losses from the resulting overuse of water.

7.9 Regulatory Agency Conflicts

This chapter's discussion of the regulation of privately owned utilities has focused on rate and service regulation. These enterprises are usually also subject to regulation by government agencies concerned with delivered water quality, general environmental impacts and conservation of water supplies. The objectives and rulings of one agency can conflict with the objectives and rulings of another agency, causing a lack of clear direction for the utility in meeting regulatory requirements. For example, Metropolitan Water Company in southern Arizona was faced with simultaneously satisfying a mandate from the Arizona Corporation Commission to meet customer demands at the regulated prices and the Arizona Department of Water Resource's system wide limits on per capita water deliveries (Arizona Water Resource, 2002). This specific conflict was later resolved by costly litigation in the Superior Court of Arizona. This case demonstrates the need for effective communication and coordination among the agencies regulating private water utilities. There is also a need for legislative bodies to be more aware of potentially conflicting laws governing water utilities and water resources.

7.10 Concluding Remarks

Since water is sold in a number of different markets, regulators and water providers must be aware of demands within each customer class as well as their total demand. While economic efficiency must be considered, water providers must also be aware of equity considerations and the need for supplies of water to be available where society feels that they are necessary.

References

Arizona Water Resources, 2002. "Ruling Fuels DWR Municipal Water Debate," *Arizona Water Resources*, 10(4) March-April.

Agthe, D.E., and R.B. Billings, 2002. "Water Price Influence on Apartment Complex Water Use," *Journal of Water Resources Planning and Management*, 128 (5): 366-369.

Babin, F.B., C.E. Willis, and P.G. Allen, 1982. "Substitution Between Water and Other Inputs," *American Journal of Agricultural Economics*, 64 (1): 149-151.

DeRooy, J., 1974. "Price Responsiveness of the Industrial Demand for Water," *Water Resources Research*, 10 (1): 403-406.

Gray, J. W., 2003. *Gray's Water Works*, unpublished manuscript, Tucson, Arizona.

Jones, C.V., and J.R. Morris, 1984. "Instrumental Price Estimates and Residential Water Demand," *Water Resources Research*, 20(2): 197-202.

Levin, R., P.R. Epstein, T.E. Ford, W. Harrington, E. Olson, and E.G. Reichard, 2002. "Drinking Water Challenges in the Twenty-First Century," *Environmental Health Perspectives*, 110 (Supplement 1) February: 43-52.

Renzetti, S., 1988. "An Econometric Study of Industrial Water Demands in British Columbia," *Water Resources Research*, 24 (10): 1569-1573.

Smith, R.S., 1998. *Profit Centers in Industrial Ecology*, Quorum Books, Westport, Connecticut.

Timmins, C.,2002. "Measuring the Dynamic Efficiency Costs of Regulator's Preferences: Urban Water Utilities in the Arid West," *Econometrica*, 70(2): 603-629.

Chapter 8

Industrial and Commercial Water Demands

Steven Renzetti
Brock University

8.1 Introduction

As the other chapters in this volume indicate, there has been a substantial amount of effort devoted by water utility managers and researchers to understand the features of urban residential households' water demands. A particular interest has been to estimate the price and income elasticities of residential water demands. It is important to note that, while households make up the largest water-using sector in most urban centres, non-residential water-use typically accounts for 30-40% of water utilities' total output (Solley, Pierce and Perlman, 1999; Environment Canada, 2002). Thus, for many water utilities facing the increasing costs of developing new sources for potable water, industrial, commercial and institutional water-use is being seen increasingly as a source for conservation through improvements in water-use efficiency.

Non-residential water-use refers to water supplied to industrial and commercial firms and institutions such as government offices, hospitals and schools. The economic features of the water demands of these businesses and organizations have not received the same attention as residential water use from researchers. This is because of the difficulties in acquiring data regarding water use in these sectors and because there is a greater degree of diversity amongst the set of non-residential water users. One can easily imagine that there are significant differences in the uses to which water is put and the factors that influence water-use across food processing plants, dentists' offices and primary schools. Nonetheless, it is worthwhile to examine the extent to which economic and engineering models of water-use can be applied profitably to further our understanding of water-use by these types of decision-makers.

D.E. Agthe et al. (eds.), Managing Urban Water Supply, 115–125.

The purpose of this chapter, then, is to briefly examine what is known regarding the economic features of water-use by industrial and commercial (I-C) firms (institutional water use is not considered here)[22]. The next section points out some of the more important characteristics of industrial and commercial water use. Of particular interest are the ways in which this sector's water-use differs from residential water use. Section 8.3 reviews the application of economic and engineering techniques to modeling and measuring I-C water-use. Section 8.4 presents several case studies that demonstrate government efforts to promote water conservation in this sector.

8.2 Features of Industrial and Commercial Water Use

There are a number of features of I-C water-use that are noteworthy. The first is the set of uses to which water is put. Manufacturing firms, typically employ water for three main reasons. Water may be applied to a production process. This includes moving intermediate inputs (coal slurry pipe line) and cleaning and as well as inclusion in final output (the production of beer). Water is used to cool intermediate inputs (the production of petroleum-based fuels) or to produce steam (thermal electric generating plants). Finally, water is used for a variety of miscellaneous purposes such as plant cleaning and personal sanitation. In addition to these manufacturing-related applications, commercial firms and institutional sites employ water in their kitchens, laundries, laboratories and even indoor fountains. One thing that can be observed from this list of water uses in the I-C sector is that there is a larger range of uses than what is typically observed in residential water use.

I-C water-use also differs in other ways from residential water use. Perhaps the most important way that it differs is that many primary and manufacturing large-scale water users are self-supplied. Indeed, the majority of manufacturing water-use in Canada and the United States is self-supplied (Solley, Pierce and Perlman, 1999; Renzetti, 2002c). The fact that large industrial water users may be self-supplied means that they do not rely upon local utilities to supply them with water or to treat their wastewater and, as a result, a provincial or state government must be concerned with monitoring and perhaps regulating their water intakes and wastewater discharges. Another way in which they differ from residential water users is that in-plant water reuse and recirculation is an important feature of water-use in many manufacturing and institutional settings. In-plant recirculation is often thought to be an important potential source for water use savings for industrial firms.

The third feature of importance is the time pattern of I-C water-use and how it differs from residential water-use (American Water Works Association, 1991). The most important difference is that I-C water-use displays less peaking on a daily or seasonal basis than residential water use, although there are exceptions such as food processing. This difference implies that I-C water use may place quite different demands on water supply and wastewater treatment systems than residential water use.

---

[22] For a more detailed discussion, the interested reader should refer to Renzetti (2002a, 2002b).

8.3 Economic Determinants of Industrial and Commercial Water Use

8.3.1 MODELING COMMERCIAL AND INDUSTRIAL WATER DEMANDS

Firms combine and transform inputs such as labor, capital, energy, materials and water into goods and services for sale. In doing so, firms must make a number of decisions related to the quantities and types of inputs and outputs and, in some cases, related to the prices of these quantities. Economists and other analysts have developed and tested sophisticated models of how firms make these decisions under a variety of circumstances. The purpose of this section is to consider the application of these models to the use of water by commercial and industrial firms.

Economists believe that profit-maximizing firms' use of any input is sensitive to its own price, the prices of other inputs and the price of output. In order to see this, suppose that a firm's technology is represented by a production function $y = f(x_{1,}\dots x_N)$ that shows that quantity of output (y) that can be produced with input quantities $(x_{1,}\dots x_N)$. Assuming that the firm faces a price of output (p) and prices of inputs $(w_{1,}\dots w_N)$ and that the firm seeks to choose output and input quantities to maximize its profits, then it can be shown that the desired quantity of any input (including intake water) may be represented in the following input demand equation:

$$x_i = h(p, w_1, \dots w_N) \quad i=1\dots N \tag{8.1}$$

Thus, the demand for intake water would be expected to be a function of its own price, the prices of all other inputs, the price of the firm's output and the firm's technology. A change in any one of these factors will alter the desired level of water use. Specifically, one would expect that increases in the price of output and decreases in the price of intake water will both increase the demand for water.[23] The effect of changes in other input prices on the demand for water will depend on whether those inputs are substitutes or complements to intake water in the firm's production process.

If a cross-price elasticity is estimated to be positive, this indicates that the two inputs are substitutes since an increase in the Economists estimate a parameter known as 'elasticity' to measure the strength and nature of these relationships. For example the own-price elasticity of intake water demand is the change in the quantity of water intake used by the firm in response to a small change in the price of intake water (both changes being expressed in percentage terms). Similarly, the cross-price elasticity of intake water demand is the change in the quantity of water intake used by the firm in response to a small change in the price of another input (both changes being expressed in percentage terms). A price increase for one (e.g. energy) leads to a greater use of the other (e.g. water).

---

[23] Until fairly recently it was commonplace to assume that the amount of water required to produce a given quantity of output was fixed (Giuliano and Spaziani, 1985). However, the model and empirical evidence presented here demonstrate that only in limited circumstances should it be expected that the ratio of water intake to output not be influenced by these factors.

There are a number of factors that influence the magnitude of the own-price elasticity of intake water demand. The first is the role of water in the firm's production process and the presence or absence of substitutes for water. In general, if water plays a very important role or has few substitutes, this will limit firms' ability to alter water use in response to higher prices. This limited ability to substitute would be reflected in relatively small price elasticity values. For example, beer producers have little choice in their water use (implying a small elasticity value) while firms that use water to cool intermediate inputs may be able to switch to air-cooling (implying a larger elasticity value). A second factor concerns the time available to adjust to price changes. A general principle in economics is that the more time a decision-maker (such as a manufacturing firm) has to adjust to a price change, the greater will be its ability to respond. Thus, in response to a significant increase in the price of intake water, a firm's short-run response may be limited to improved housekeeping practices and minor repairs. Over time, however, the firm may be able to implement more significant changes such as a switch from once-through systems to in-plant recirculation and other process changes that require capital investments. A third factor is the importance of the input to the firm's costs of production. It is generally accepted that the larger the share in costs for a specific input, the more responsive a firm will be to changes in its price. In the case of water (whose cost share rarely exceeds one percent for commercial and industrial users), this would suggest that price changes would have to be quite significant before they elicit major reactions from users.

The next step in understanding the factors that influence I-C water-use is to gather data and estimate or calibrate the demand model specified above. There are two general approaches that can be taken to do this. The first approach applies statistical techniques to estimate a regression equation that indicates the relative strengths of the various factors identified above in determining I-C water use. These econometric methods have the advantage of being able to characterize water-use over a variety of time periods and types of production processes. The second approach is to use mathematical programming methods to understand I-C water use. Programming models typically are used to represent water-use in specific production processes. As such, they require large amounts of detailed information regarding the production process.

8.3.2 EMPIRICAL ESTIMATES[24]

A number of researchers have considered the structure of industrial water demands. Early statistical efforts fitted linear demand equations to plant-level data which included observations on the price of water, the quantity of intake water and level and type of output.[25] For example, De Rooy (1974) finds that price and output elasticity estimates

[24] This sub-section draws on Renzetti (2002c).

[25] An important issue in these, and subsequent studies, concerns the specification of the price of intake water. In the case of publicly supplied firms, industrial water prices are typically a combination of a connection fee and a complex rate schedule (Dinar and Subramanian, 1997). When price is a function of quantity this creates the possibility of introducing what is known as a simultaneity bias into the estimated demand equation. Ziegler and Bell (1984) and Renzetti (1992) address this issue by estimating demand equations using instrumental variable techniques.

vary according to the use of the intake water: -0.894 and 1.212 (cooling); -0.354 and 1.359 (process); -0.590 and 1.243 (steam generation). The output elasticity of input demand measures the percentage change in the use of an input (such as intake water) when output increases by a small percentage. Thus, the estimated output elasticity of 1.212 indicates that, when the quantity of output rises by 1%, the firm's use of intake water for cooling purposes is predicted to increase by 1.212%. Therefore, in each case cited, water use increases more than in proportion to output.

Subsequently, researchers such as Babin, Willis and Allen (1982) and Renzetti (1992) have used a systems-of-equations approach to estimate the industrial demand for water. This increases the efficiency of estimation and allows the authors to examine the relationship between water and other inputs. Babin, Willis and Allen find that the own price elasticity of intake water ranges from 0.0 to –0.66 with an estimate of –0.66 for all sectors combined. Water's relationship with other inputs also varies across sectors. Water is a substitute for capital in some sectors and is a complement in others. Renzetti (1992) finds that water intake's price elasticities range from –0.153 to –0.589 depending on the sector. Water recirculation is seen to be a substitute for intake for all industries. Industries that are heavy water users (Paper and Wood products, Chemicals, Petroleum, Food and Beverage) also are the most sensitive to price changes.

More recently, Dupont and Renzetti (2001) incorporated water recirculation and technological change in their estimation model of water use in Canadian manufacturing firms during 1981-1991. Water intake is found to be a substitute for energy, labor, capital and water recirculation. The relationship between water intake and recirculation is stronger when water-use is process-related rather than related to cooling and steam production. Finally, because the estimation employs time series data, the authors are able to determine that technological change was biased during that time period in the direction of increased water intake and decreased water recirculation.

The work of Dupont and Renzetti (2001) points out a challenge that faces researchers when they wish to use econometric models to estimate the water demands of self-supplied industrial water users. When firms are self-supplied, there may be no external price for intake water although these firms do still incur several types of cost associated with their water use. These costs can include internal costs resulting from pumping, treatment prior to use (chlorination, descaling etc.) and external costs arising from fees and charges set by government. In addition, firms may incur internal costs associated with water disposal such as treatment, removal of effluents, pumping and on-site storage. It is very likely that information regarding these costs will not be publicly available and firms may not even collect this information themselves. This means that it is difficult to specify and measure the cost associated with an incremental increase in water intake. Renzetti (1992) addresses this problem by relying on plant-level observations on water utility prices, government water charges and internal water-use costs to construct a proxy for the price of water.

An alternative to using econometric methods to model water demands is to use programming techniques. The solutions to linear programming problems have several interesting features that differentiate them from the estimated econometric demand models discussed above. For example, input demand equations derived from linear programming models are typically piece-wise continuous and linear. Discontinuities in the demand equation correspond to the adoption of new production processes (such as the installation of a cooling tower that permits recirculation and re-use of cooling water). Simulations conducted by Calloway, Schwartz and Thompson (1974) demonstrate the significance of this feature. Their results indicate that for intake water prices anywhere below 3.2 cents (1971 U.S.$) per 1000 gallons (0.85cents/m^3), a plant producing ammonia employs a water intensive 'once-through' cooling system. At 3.2 cents per 1000 gallons however, the plant switches to recycling its cooling water and this reduces water intake by 95% and further price increases do little to encourage further water conservation. In contrast, econometric models tend to predict smaller responses to prices changes because they assume firms are able to make incremental adjustments to their water use.

It is important to remember that firms that use water in their production processes must find some way to dispose of that water. Water quality regulations typically impose limits on the quantity and/or concentration of effluents that can be discharged. In fact, some analysts argue that water quality regulations have been a driving force behind observed reductions in industrial water intake in the U.S. and Europe (Solley, Pierce and Perlman, 1999). Empirical evidence also suggests that economic factors may influence industrial water discharge decisions. For example, Renzetti (1992) finds that, for the Canadian manufacturing sector as a whole, the price elasticity of water discharge is -0.97, which means that for a 10% increase in disposal fees firms would cut water use by 9.7%. This parameter estimate suggests that firms' wastewater disposal decisions are fairly responsive to unit costs and could be reduced through a greater reliance on enhanced sewerage charges.

The discussion, thus far, has concentrated on water-use in the manufacturing sector. A small number of research studies have considered commercial and institutional water use. The work done in this area suggests that this sector's water-use is somewhat sensitive to water prices and the firm's level of output although the strength of these factors is not as great as that seen in residential and industrial water use. Schneider and Whitlatch (1991), for example, estimate price and output elasticities for several municipally supplied user groups including commercial, industrial and government. The account-specific price elasticities (following the order of user groups set out above) are -0.918, -0.438 and -0.781. Given the diversity of water users in these sectors, providing general conclusions regarding the nature of their water-use is quite difficult. An alternative approach to studying water-use in these sectors is to develop Best Management Practices that establish benchmark levels of water-use for different types of water-use (kitchens, laundry, bathrooms, outdoor uses, etc.). As indicated in section 8.4 below, this approach has proven popular with governments who are interested in promoting water conservation in these sectors.

8.3.3 THE VALUE OF COMMERCIAL AND INDUSTRIAL WATER USE

The use of water, like any other productive input, provides value to a firm. While researchers have developed a number of methods to estimate these values there are few studies that actually provide estimates of the value of water to commercial enterprises. This stands in contrast to the extensive amount of research conducted into households' valuation of water (Renzetti, 2002a, chapters 8 and 9). The reasons for this lack of research include the scarcity of data relating to industrial water-use and the fact that it is rare for water to account for more than one percent of firms' total costs (Dupont and Renzetti, 2001; United Kingdom Environment Agency, 2001).

One valuation method relies upon information regarding the incremental costs of internal water recirculation to derive an estimate of the value of intake water. The reasoning behind this approach is that the incremental cost of recirculation should provide an upper bound on the amount a firm would be willing to pay to acquire additional intake water (assuming that there is no difference in the quality of the two types of water). With respect to water use in cooling (such as petroleum refineries and electrical generation plants), Gibbons (1986) reports an estimate of the unit cost of moving from once-through to evaporative cooling to be small: $6-10/acre-foot (in 1980 dollars). With respect to water use in process applications, values are higher because water recirculation is more costly due to the need to address water quality changes. Gibbons reports values ranging from $16 to $75/acre-foot for process water applications with even higher values in cases where water use implies significant reductions in water quality.

The residual imputation approach estimates the value of water by subtracting the costs of all of a firm's non-water inputs from that firm's revenues. What remains (the 'residual') can be argued to be that portion of total revenues that is attributable to the productivity of water and, hence, an estimate of its value to the firm. While this approach appears attractive in principle, water's relatively small cost share and the difficulty of costing inputs such as capital[26] and good will present significant challenges implementing it. Moore (1999) provides an example of applying this procedure for the water used in irrigation.

In the case where there is sufficient data available to estimate a water intake demand equation, the firm's valuation (or willingness to pay) for water could be inferred directly from the height of the demand equation for any given quantity of water. Wang and Lall (1999) was an effort to apply this approach in order to estimate the value of water using a sample of Chinese manufacturing firms.

In summary, despite the availability of a number of approaches to value water-use in the I-C sector, data limitations inhibit their application and this remains an understudied area. Frederick, VandenBerg and Hanson (1997) provide a valuable compilation of available estimates of the value of water in the U.S. In the case of industrial applications, the authors find a range of average values from $28 to $802 with a median value of $132 (1994 U.S. $/acre-foot).

8.4 Case Studies

As indicated at the beginning of this chapter, I-C water-use is an important issue for many water utilities. While industrial water use is declining throughout North America,

[26] The value of capital is the opportunity cost of its current utilization, or what it could earn in its next most valuable use. In practice it is very difficult to determine the correct value.

large individual industrial and commercial water users can place significant pressures on local water systems. One particularly interesting example of this type of situation is occurring in areas such as Hawaii, Florida and southern Europe where the growth in tourism resorts and hotels is placing a strain on the capacity of local infrastructure (Sinclair and Stabler, 1997).

These kinds of pressures have led a number of local, state and national governments to explore the possibilities of water conservation in the I-C sector. The case studies that follow provide examples of the approaches that have been taken. It is interesting to note that, despite research showing industrial (and, to a lesser extent, commercial) water-use to be sensitive to prices, a number of governments have embarked on fairly extensive programs to assist firms to reduce their water use. A lower cost alternative for governments and local utilities may be to rely more heavily on enhanced price signals (e.g. implementing peak-load pricing to inform users of the costs of water-use at different times in the year) to promote water conservation.

8.4.1 CALIFORNIA'S URBAN CONSERVATION OFFICE

The State of California has a long history of managing its water resources. Recent experience with drought-induced reductions in water supplies and growing conflicts amongst water users provided the state government with a strong incentive to examine water conservation as part of its overall management strategy. As a result, the State of California's Department of Water Resources established an Urban Conservation Office. This agency provides a variety of types of information about how to reduce urban water use-including commercial and industrial water use (California Department of Water Resources, 1994; Vickers, 2001).

In addition, the California Urban Water Conservation Council is a state body that is mandated to implement conservation plans that reflect "Best Management Practices" in Californian municipal water utilities. These conservation plans will include the following (CUWCC, 2002):

- A ranking of commercial and industrial water users according to their water-use levels.
- Development of conservation plans to reduce water-use by these sectors. The conservation plan must either (a) contain a plan to implement a commercial, industrial and institutional water-use survey and customer incentive program or (b) demonstrate that it will achieve water-use reductions in the commercial, industrial and institutional sector equaling or exceeding targets described in State guidelines.

A number of other American States are implementing similar programs. For example, North Carolina has developed an extensive directory of water conservation methods for industrial and commercial firms to examine and a set of guidelines for firms' evaluation of their economic feasibility (North Carolina Department of Environment and Natural Resources (1998).

8.4.2 ONTARIO'S GREEN INDUSTRIAL ANALYSIS PROJECT

The Ontario Ministry of Environment and Energy (OMOEE) in Canada implemented a program to encourage 'partnership in pollution prevention and resource conservation' (Ontario Ministry of the Environment and Energy, 1997). Under this program industrial companies located in Ontario may seek Ministry services that will help them achieve a variety of environmental goals. These goals include a more efficient use of energy and water. The following are examples of firms that have been successful in reducing their use of water:

- When faced with a sewer surcharge on effluents Hostess-Frito Lay (a food processor) introduced a change in its processing operations that resulted in a reduction of water intake of 24,000 cubic metres per year. For a one time capital cost of $2 million the company found itself able to save $0.6 million annually in effluent charges, thereby indicating a payback period on the new technology of about 3 years.
- QUNO Corporation (a newsprint mill) was a very heavy user of both water and electricity inputs. Its goal was to reduce its usage of both of these inputs in order to save money. It spent $5.5 million to make improvements to its water pumps and to increase the amount of water recirculation. Not only was it able to reduce its use of the two inputs, it also found itself saving $1.4 million annually.
- When Rieter Automotive North America Inc. (an automobile parts manufacturer located in Tillsonburg, Ontario, with a population of 85,000) found itself in the situation of using 25 % of the town's entire water supply it set two goals. The first was to reduce its water intake. The second was to increase water recirculation. By recirculating its cooling water the company not only reduced its water intake by 200,000m^3 annually, it also saved $0.12 million annually from an investment of $0.2 million.

8.4.3 U.K. ENVIRONMENT AGENCY'S "WATERWISE" PROGRAM

The United Kingdom is not usually thought of as a country threatened by water shortages. However, a prolonged drought in the early 1990's and regional conflicts amongst water utilities, industrial water users and environmental interests prompted regulators to consider more carefully the allocation of the U.K.'s water resources. As part of this process, the government of the United Kingdom established the National Water Demand Management Centre (NWDMC) as a clearinghouse for information related to water conservation.

A key part of the mandate of the NWDMC is to facilitate water conservation in major water using sectors. Specifically, the Waterwise initiative identifies Best Management Practices (BMP) for a wide range of commercial and industrial water applications (United Kingdom Environment Agency, 2001). These BMPs can be used to assess water-use in a specific plant or office and form the basis for a firm's water management plan. The NWDMC points out that such a plan should have the following steps:

- Obtain management and staff support for water conservation
- Identify the full cost of water-use(including treatment costs, maintenance costs for the water supply network, disposal costs and staff training costs)
- Quantify current water uses
- Re-assess current water-use through reference to use levels specified in the BMP guidelines
- Identify and evaluate possible water conservation measures. Evaluation includes calculating pay-back periods for any investments.
- Implement and monitor water conservation efforts.

According to the NWDMC, undertaking this type of effort can have significant effects: "Considerable reductions in costs can be achieved (potentially up to 80 per cent where no previous action has been taken). Many actions to reduce water-use have short payback periods of 6 to 24 months" (United Kingdom Environment Agency, 2001, p.8).

8.5 Conclusions

For many cities, water-use by industrial plants, retail businesses, hotels and other firms constitutes an important share in the total demands placed on water supply and wastewater treatment infrastructures. However, it is only relatively recently that economists have begun to share engineers' interest in the role water plays in industrial and commercial processes. Unfortunately, the degree to which the various facets of I-C water-use are sensitive to price changes and government non-price water conservation programs has not been as well established as in the case of water use in other sectors.
Nonetheless, the limited amount of evidence suggests that most facets of industrial and commercial water use are responsive to economic forces.

References

American Water Works Association, 1991. *Manual of Water Supply Practices: Water Rates* 4th ed. Denver: Colorado.

Babin F., C. Willis and P. Allen, 1982. "Estimation of Substitution Possibilities between Water and Other Production Inputs," *American Journal of Agricultural Economics*, 64(1): 148-151.

California Department of Water Resources, 1994. *Water Efficiency Guide for Business Managers and Facility Engineers*, Sacramento, California.

Calloway, J., A. Schwartz and R. Thompson, 1974. "Industrial Economic Model of Water-use and Waste Treatment for Ammonia," *Water Resources Research* 10(4): 650-658.

CUWCC, 2002. "Implementation of Best Management Practices" website (accessed May 18, 2002).

De Rooy, J., 1974. "Price Responsiveness of the Industrial Demand for Water" *Water Resources Research* 10(3): 403-406.

Dinar, A. and A. Subramanian, eds., 1997. *Water Pricing Experiences: An International Perspective*, World Bank Technical Paper No. 386, The World Bank, Washington, D.C.

Dupont, D. and S. Renzetti, 2001. "Water's Role in Manufacturing," *Environmental and Resource Economics* 18(4): 411-432.

Environment Canada, 2002. "Freshwater" website (accessed May 18, 2002).

Frederick, K., T. VandenBerg and J. Hanson, 1997. *Economic Values of Freshwater in the United States*, Discussion Paper 97-03, Resources for the Future, Washington D.C.

Gibbons, D., 1986. *The Economic Value of Water*, Resources for the Future, Washington, D.C.

Giuliano, G. and F. Spaziani, 1985. "Water-use statistics in industry. Experiences from regional surveys and planning studies in Italy," *Statistical Journal of the United Nations*, 3: 229-235.

Moore, M., 1999. "Estimating Irrigators' Ability to Pay for Reclamation Water," *Land Economics* 75(4): 562-578.

North Carolina Department of Environment and Natural Resources, 1998. *Water Efficiency Manual for Commercial, Industrial and Institutional Facilities*, Raleigh, North Carolina.

Ontario Ministry of the Environment and Energy, 1997. *Project Profiles: Green Industrial Analysis*, Industry Conservation Branch, Toronto.

Renzetti, S., 1992. "Estimating the Structure of Industrial Water Demands: The Case of Canadian Manufacturing," *Land Economics* 68(4): 396-404.

Renzetti, S., 2002a. *The Economics of Water Demands*, Kluwer Academic Press, Norwell, Massachusetts.

Renzetti, S. ed., 2002b. *The Economics of Industrial Water Use*, Edward Elgar. Cheltenham, UK.

Renzetti, S., 2002c. "Economics of Industrial Water Demands" in J.H. Lehr, J. Keeley and J. Lehr eds. *The Encyclopaedia of Water*, John Wiley, forthcoming.

Schneider, M. and E. Whitlatch, 1991. "User-Specific Water Demand Elasticities," *Journal of Water Resources Planning and Management* 117(1): 52-73.

Sinclair, M. and M. Stabler, 1997. *The Economics of Tourism*, Routledge. London.

Solley, W., R. Pierce and H. Perlman, 1999. *Estimated Use of Water in the United States in 1995* United States Geological Survey Circular 1200.

United Kingdom Environment Agency, 2001. *Waterwise: Good for Business and Good for the Environment*, The National Water Demand Management Centre, London.

Vickers, A., 2001. *Handbook of Water Use and Conservation*, Californian Urban Water Conservation Council, Los Angeles, California.

Wang, H. and S. Lall, 1999. "Valuing Water for Industry: A Marginal Productivity Approach," *Sinosphere* 2(3): 27-50.

Ziegler J. and S. Bell, 1984. "Estimating Demand by Self-Supplied Firms," *Water Resources Research* 20(1): 4-8.

Chapter 9

Public Drinking Water in the United States

Ronnie Levin
Harvard School of Public Health

The provision of safe and reliable drinking water to most of the U.S. population is ranked as one of the 10 major public health achievements of the twentieth century (CDC, 1999). Over 250 million Americans, who are about 90% of the estimated population, currently are served by about 54,000 public water systems. In those systems, over 80 contaminants are regulated and hundreds of water quality parameters are monitored. Total expenditures for public drinking water in the U.S. are about $36 billion per year.

9.1 Regulation of U.S. Public Drinking Water

9.1.1 HISTORY OF DRINKING WATER REGULATION

Archeological evidence shows that irrigation was known in the ancient Near East, and an early Sanskrit text discusses the purification of drinking water by boiling and filtering in India (Baker, 1949). Before 2000 BCE, the rulers of Babylonia and Egypt constructed canal and dam systems to impound the Euphrates and Nile Rivers, which controlled flooding, provided irrigation water and also supplied water for domestic use. Both the ancient Greeks and especially the Romans addressed obtaining and maintaining a reliable supply of good water for their cities. The Romans, indeed, constructed a vast system of aqueducts to bring the clean waters of the Apennine Mountains into Rome, and built settling basins and filters along these mains to ensure the clarity of the water. The Roman water supply and particularly the wastewater disposal components were so

D.E. Agthe et al. (eds.), Managing Urban Water Supply, 127–145.

well designed and built that most still stand today; in fact, the Great Sewer of Rome, the Cloaca Maxima, is still in service (Fair and Geyer, 1958).

Compared to these early efforts, further developments in the hygienic control of water supply and waste water disposal occurred quite recently. Through the 1850s, it was commonly thought that disease was caused by foul air. Only in 1854 did John Snow show that cholera was waterborne by taking the handle off the Broad Street (London) water pump during an epidemic and arresting the outbreak. John Snow could only speculate about a 'cholera poison' in the water; *Cholera vibrio* was not yet identified. (Robert Koch identified the pathogen in 1883.)

Thereafter, the regulation of water based on health concerns developed relatively rapidly. In London, legislation required that all London water companies filter their water by 1857. In the U.S., Federal authority to establish standards for drinking water systems began later, in 1893, with the Interstate Quarantine Act, which authorized the Surgeon General of the U.S. Public Health Service "to make and enforce such regulations as in his judgment are necessary to prevent the introduction, transmission, or spread of communicable disease from foreign countries into the states or possessions, or from one state or possession into any other" (McDermott, 1973). The 1893 Act led to the promulgation of the Interstate Quarantine Regulations of 1894, which in turn led to the first U.S. water-related regulation: prohibiting the use of a common drinking cup on interstate carriers.

Because safe drinking water began in the U.S. as an interstate issue, its regulation resided with the U.S. Treasury Department. In 1914, the U.S. Treasury Department with the U.S. Public Health Service (PHS) issued the first official drinking water standard - for coliform bacteria, which applied to "interstate carrier conveyances" such as trains (Taylor, 1977). In 1925, new regulations were issued, revising the earlier coliform standard and adding new sections on water source and supply, and physical and chemical characteristics (such as lead, copper, and zinc). Another revision was made in 1942, increasing the number of samples collected, calling for sampling within the distribution system (not at the treatment plant), revising some of the existing standards and establishing some new ones (Larson, 1989). Revisions and additions to these standards also occurred in 1946 and in 1962 (McDermott, 1973), and in 1963, the PHS established a Drinking Water Advisory Committee on the use of the Public Health Service Drinking Water Standards. The Drinking Water Advisory Committee was acutely aware of the constraint on establishing national standards, especially those that did not specifically address waterborne disease, and issued important reports in 1967 and 1971 calling for revisions to both the substance and authority for drinking water regulation in the U.S. (Larson, 1989).

States and local authorities could still either adopt or reject the federal standards for their own stationary public drinking water systems. By the early 1970s, in fact, most states had not adopted the federal standards (Levin et al; 2002). The U.S. Congress passed the Safe Drinking Water Act (SDWA) in 1974, which gave the newly created Environmental Protection Agency (EPA) authority to regulate all U.S. public drinking water, and required that all states adopt the federal standards or individual ones that were at least as stringent as the federal standards. Over the next decade or so, EPA ratified most of the approximately 30 older PHS guidelines as "interim standards" and adopted one new one in 1979 - for trihalomethanes (THMs); these standards were then enforceable on all public water systems in the U.S.

The U.S. Congress, frustrated by the slow pace of improvement, significantly revised the SDWA in 1986 and required EPA to regulate 83 specific contaminants in the next decade, almost triple EPA's activity in the previous 12 years. EPA was also required to regulate 25 new contaminants every three years thereafter. This generated a backlash supported by many water utilities and state and local government officials to relax the 1986 revisions of the SDWA. Simultaneously, however, public concern about drinking water safety was increasing, fueled by the 1993 waterborne disease outbreak in Milwaukee that sickened 400,000 and resulted in up to 100 deaths (MacKenzie et al., 1994; Moxie et al., 1997) and studies that showed health risks associated with exposure to some common drinking water contaminants.

In 1996, Congress revised the SDWA again. Major revisions included a new emphasis on community "right to know", such as mandatory annual reports on water quality by each water utility, plus a concern for vulnerable populations, such as children, pregnant women and the chronically ill, in setting new drinking water standards. On the other hand, the requirement to regulate 25 new contaminants every three years was replaced by a mandate to *consider* regulating at least five new contaminants every five years. In addition, the 1996 amendments allowed consideration of costs and benefits, and risk-risk trade-offs in setting drinking water standards. And EPA was required to review and, where feasible, strengthen all drinking water standards every six years.

A recent investigation of all state and federal laws protecting the public against waterborne disease and other water quality problems in the U.S. showed serious gaps and discrepancies (Gostin et al., 2000). The authors observe that, while our federalist legal system welcomes state experimentation, in fact the current system hinders public health efforts to prevent illness caused by drinking water and to control waterborne disease.

9.1.2 DEFINITIONS OF PUBLIC WATER SYSTEMS IN THE U.S.

Under the US Safe Drinking Water Act of 1974, EPA is given authority to regulate *public water systems*, which are defined statutorily as 'systems that have at least 15 service connections or regularly serve an average of at least 25 people daily for at least 60 days of the year'. Drinking water systems serving less than 25 people or having fewer than 15 service connections are considered 'private water systems,' and are not covered by the SDWA.

Public water systems (PWSs) are divided into three groups for regulatory purposes. PWSs that serve year-round residents are called *community water systems* (CWSs), while those serving nonresidential areas or customers, such as restaurants, motels and campgrounds are called *noncommunity water systems* (NCWSs). Finally, *nontransient noncommunity systems* (NTNWSs) are CWSs that regularly serve the same *non*residential customers, such as factories or day schools that operate their own drinking water systems. Figures 9.1and 9.2 show the distribution of all PWSs in the United States. Less than one-third of the 190,000 *public* water suppliers are *community*

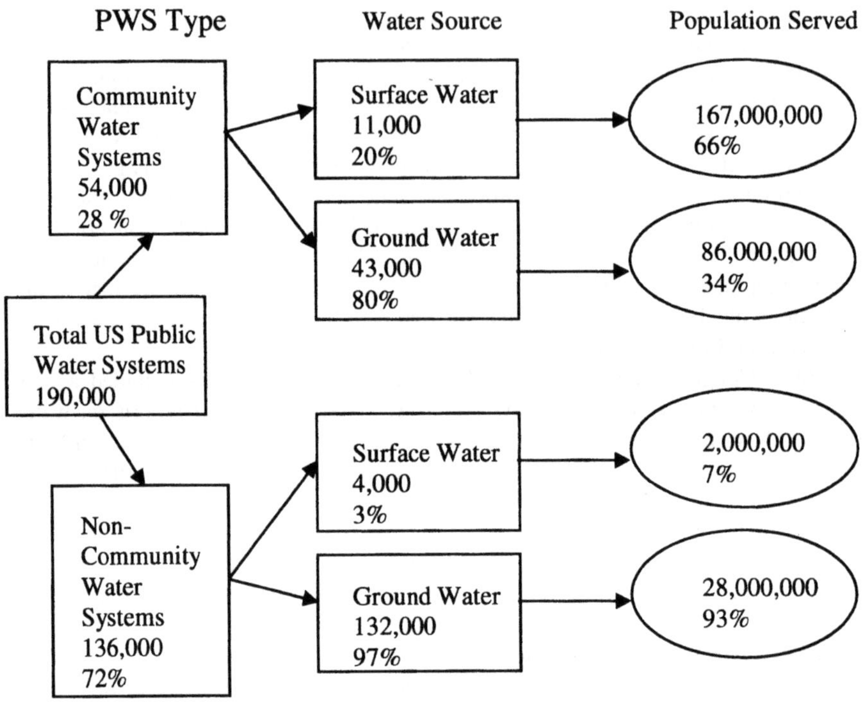

Figure 9.1 Public Water Systems by Type and Water Source, FY 2000

Note: There is some double counting in the estimation of population served, especially for the non-community water systems because some people served by NCWSs at school, work, camp, etc. live in residences that are connected to CWSs. Source: US EPA, 2000. Schematic updated from Karalekas and Trax, 1989.

water systems, while the 11,000 community water systems utilizing surface water supplies as a source serve over half of the U.S. population of 283 million.

9.1.3 CURRENT U.S. DRINKING WATER STANDARDS

The U.S. Safe Drinking Water Act (SDWA), as amended in 1986 (1986 SDWA), requires EPA to publish a "maximum contaminant level goal" (MCLG) for each contaminant which, in the judgment of the EPA Administrator, "may have any adverse effect on the health of persons and which is known or anticipated to occur in public water systems" [Section 1412(b)(3)(A)]. MCLGs are to be set at a level at which "no known or anticipated adverse effect on the health of persons occurs and which allows an adequate margin of safety." (Section 1412(b)(4)) The 1996 amendments to the SDWA revised some of the citations and also augmented the Act with additional language, including the specification that EPA consider the effects of contaminants upon sensitive subpopulations, such as infants, children, pregnant women, the elderly and those with a history of serious illness.

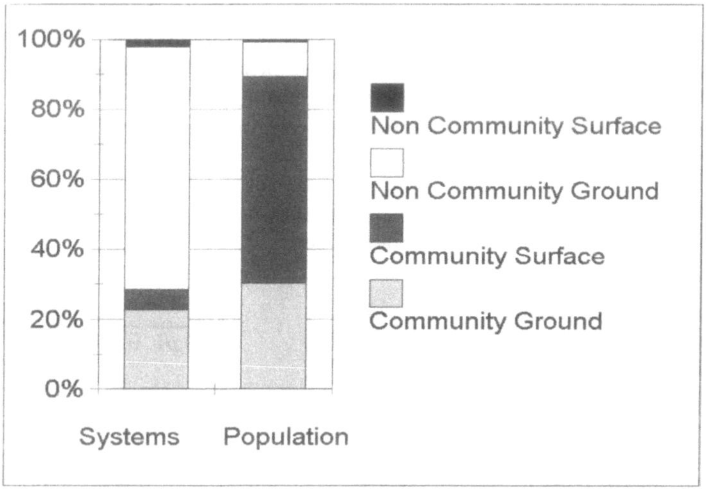

Figure 9.2 Percentage Distribution of Public Water Systems, FY 2000
Source: US EPA, 2000. See note to Figure 9.1.

The SDWA requires that at the same time that EPA publishes an MCLG, which is a non-enforceable health goal, it also must publish a *National* Primary Drinking Water Regulation (NPDWR) that specifies either a maximum contaminant level (MCL) or a treatment technique, both of which are enforceable [Section 1401(1) and Section 1412(a)(3)]. EPA can establish a treatment technique in lieu of an MCL when the Agency finds that "it is not economically or technologically feasible to ascertain the level of the contaminant." The MCL must be set as close to the MCLG as is feasible, which is defined as "with the use of the best technology, treatment techniques and other means, which the Administrator [of EPA] finds, after examination for efficacy under field conditions and not solely under laboratory conditions, are available (taking cost into consideration)." Costs are defined as those estimated for large water systems. The MCLG for all carcinogens must be set at 'zero'. Tables 9.1a through 9.1f contain a list of all the current NPDWRs.

The 1996 SDWA also required the EPA to establish criteria for a monitoring program for unregulated contaminants and to publish a list of contaminants to be evaluated for future drinking water regulations. The data generated by the current Unregulated Contaminant Monitoring Rule (UCMR) will be used to evaluate and prioritize contaminants on the Drinking Water Contaminant Candidate List, a list of contaminants EPA is considering for possible new drinking water standards. The following EPA definitions are used throughout Table 9.1:

- Maximum Contaminant Level (MCL) - the highest level of a contaminant that is allowed in drinking water. MCLs are set as close to MCLGs as feasible using the best available treatment technology and taking cost into consideration. MCLs are enforceable standards.

- Maximum Contaminant Level Goal (MCLG) - The level of a contaminant in drinking water below which there is no known or expected risk to health. MCLGs allow for a margin of safety and are non-enforceable public health goals.
- Treatment Technique (TT) - A required process intended to reduce the level of a contaminant in drinking water.

Table 9.1a U.S. National Primary Drinking Water Regulations
Microorganisms

Microorganisms	**MCLG** (Mg/L)	**MCL** (Mg/L) or **TT**	**Notes** Mg/L = milligrams/liter = parts/million
Cryptosporidium	Zero	TT[a]	
Giardia lamblia	Zero	TT[a]	
Heterotrophic plate count	n/a	TT[a]	HPC has no health effects; it is an analytic method used to measure the variety of bacteria that occur in water.
Legionella	zero	TT[a]	
Total Coliforms (including fecal coliform and E. Coli)	Zero	5.0%[b]	Not a health threat in itself; it is used to indicate whether other potentially harmful bacteria may be present. Coliforms are naturally present in the environment, as well as in feces; fecal coliforms and E. coli only come from human and animal fecal waste.
Turbidity	Na	TT[a]	Turbidity is a measure of the cloudiness of water. It is used to indicate water quality and filtration effectiveness. Higher turbidity levels are often associated with higher levels of disease-causing microorganisms such as viruses, parasites and some bacteria. These organisms can cause illness.
Viruses (enteric)	Zero	TT[a]	

Source for all parts of Table 9.1: http://www.epa.gov/safewater/mcl.html See footnotes on following page.

Table 9.1b U.S. National Primary Drinking Water Regulations
Disinfection Byproducts

Disinfection Byproducts	**MCLG** (Mg/L)	**MCL** (Mg/L) or **TT**
Bromate	Zero	0.010
Chlorite	0.8	1.0
Total Trihalomethanes (TTHMs)	none[a]	0.080
Haloacetic acids (HAA5)	na[a]	0.060

[a] Although there is no collective MCLG for this contaminant group, there are individual MCLGs for some of the individual contaminants: **Trihalomethanes**: bromodichloromethane (zero); bromoform (zero); dibromochloromethane (0.06 mg/L). Chloroform is regulated with this group but has no MCLG. **Haloacetic acids**: dichloroacetic acid (zero); trichloroacetic acid (0.3 mg/L). Monochloroaceticacid, bromoacetic acid, and dibromoacetic acid are regulated with this group but have no MCLGs.

Footnotes to Table 9.1 a

[a] EPA's surface water treatment rules require systems using surface water or ground water under the direct influence of surface water to (1) disinfect their water, and (2) filter their water or meet criteria for avoiding filtration so that the following contaminants are controlled at the following levels: **Cryptosporidium** (as of1/1/02 for systems serving '10,000 and 1/14/05 for systems serving <10,000) 99% removal. **Giardia lamblia**: 99.9% removal/inactivation

Viruses: 99.99% removal/inactivation. **Legionella**: No limit, but EPA believes that if Giardia and viruses are removed/ inactivated, Legionella will also be controlled. **Turbidity**: At no time can turbidity (cloudiness of water) go above 5 nephelolometric turbidity units (NTU); systems that filter must ensure that the turbidity go no higher than 1 NTU (0.5 NTU for conventional or direct filtration) in at least 95% of the daily samples in any month. As of January 1, 2002, turbidity may never exceed 1 NTU, and must not exceed 0.3 NTU in 95% of daily samples in any month.

Heterotrophic Plate Count: No more than 500 bacterial colonies per milliliter. **Long Term 1 Enhanced Surface Water Treatment** (Effective Date: January 14, 2005); Surface water systems or ground water systems under the direct influence of surface waters serving fewer than 10,000 people must comply with the applicable Long Term 1 Enhanced Surface Water Treatment Rule provisions (e.g. turbidity standards, individual filter monitoring, Cryptosporidium removal requirements, updated watershed control requirements for unfiltered systems). **Filter Backwash Recycling**; The Filter Backwash Recycling Rule requires systems that recycle to return specific recycle flows through all processes of the system's existing conventional or direct filtration system or at an alternate location approved by the state.

[b] More than 5.0% samples total coliform-positive in a month. (For water systems that collect fewer than 40 routine samples per month, no more than one sample can be total coliform-positive per month.) Every sample that has total coliform must be analyzed for either fecal coliforms or E. coli. If two consecutive TC-positive samples, and one is also positive for E.coli fecal coliforms, system has an acute MCL violation.

Table 9.1c U.S. National Primary Drinking Water Regulations
Disinfectants

Disinfectants	**MRDLG**[a] (Mg/L)	**MRDL**[a] (Mg/L)
Chloramines (as Cl2)	4.0	4.0
Chlorine (as Cl2)	4.0	4.0
Chlorine dioxide (as ClO2)	0.8	0.8

[a] **Maximum Residual Disinfectant Level (MRDL)** - The highest level of a disinfectant allowed in drinking water. There is convincing evidence that addition of a disinfectant is necessary for control of microbial contaminants. **Maximum Residual Disinfectant Level Goal (MRDLG)** - The level of a drinking water disinfectant below which there is no known or expected risk to health. MRDLGs do not reflect the benefits of the use of disinfectants to control microbial contaminants.

Table 9.1d U.S. National Primary Drinking Water Regulations
Inorganic Chemicals

Inorganic Chemicals	**MCLG** (Mg/L)	**MCL** (Mg/L) or **TT**
Arsenic	zero	0.010 as of 01/23/06
Asbestos (fiber '10 micrometers)	7 million fibers per liter (MFL)	7 MFL
Barium	2	2
Beryllium	0.004	0.004
Cadmium	0.005	0.005
Chromium (total)	0.1	0.1
Copper	1.3	TT[a] Action Level = 1.3
Cyanide (as free cyanide)	0.2	0.2
Fluoride	4.0	4.0
Lead	zero	TT[a] action level = 0.015
Mercury (inorganic)	0.002	0.002
Nitrate (measured as Nitrogen)	10	10

Inorganic Chemicals	**MCLG** (Mg/L)	**MCL** (Mg/L) or **TT**
Nitrite (measured as Nitrogen)	10	10
Selenium	0.05	0.05
Thallium	0.0005	0.002

[a] **Lead and copper** are regulated by a Treatment Technique that requires systems to control the corrosiveness of their water. If more than 10% of tap water samples exceed the action level, water systems must take additional steps. For copper, the action level is 1.3 mg/L, and for lead is 0.015 mg/L.

Table 9.1e U.S. National Primary Drinking Water Regulations
Organic Chemicals

Organic Chemicals	**MCLG** (Mg/L)	**MCL** (Mg/L) or **TT**
Acrylamide	zero	TT[a]
Alachlor	zero	0.002
Altrazine	0.003	0.003
Benzene	zero	0.005
Benzo(a)pyrene (PAHs)	zero	0.0002
Carbofuran	0.04	0.04
Carbon tetrachloride	zero	0.005
Chlordane	zero	0.002
Chlorobenzene	0.1	0.1
2,4-D	0.07	0.07
Dalapon	0.2	0.2
1,2-Dibromo-3-chloropropane (DBCP)	zero	0.0002
o-Dichlorobenzene	0.6	0.6
p-Dichlorobenzene	0.075	0.075
1,2-Dichloroethane	zero	0.005

Organic Chemicals	**MCLG** (Mg/L)	**MCL** (Mg/L) or **TT**
1,1-Dichloroethylene	0.007	0.007
cis-1,2-Dichloroethylene	0.07	0.07
trans-1,2-Dichloroethylene	0.1	0.1
Dichloromethane	zero	0.005
1,2-Dichloropropane	zero	0.005
Di(2-ethylhexyl) adipate	0.4	0.4
Di(2-ethylhexyl) phthalate	zero	0.006
Dioxin (2,3,7,8-TCDD)	zero	0.00000003
Diquat	0.02	0.02
Endothall	0.1	0.1
Endrin	0.002	0.002
Epichlorohydrin	zero	TT[a]
Ethylbenzene	0.7	0.7
Ethylene dibromide	zero	0.00005
Glyphosate	0.7	0.7
Heptachlor	zero	0.0004
Heptachlor epoxide	zero	0.0002
Hexachlorobenzene	zero	0.001
Hexachlorocyclopentadiene	0.05	0.05
Lindane	0.0002	0.0002
Methoxychlor	0.04	0.04
Oxamyl (Vydate)	0.2	0.2
Polychlorinated biphenyls (PCBs)	zero	0.0005

Organic Chemicals	**MCLG** (Mg/L)	**MCL** (Mg/L) or **TT**
Pentachlorophenol	zero	0.001
Picloram	0.5	0.5
Simazine	0.004	0.004
Styrene	0.1	0.1
Tetrachloroethylene	zero	0.005
Toluene	1	1
Toxaphene	zero	0.003
2,4,5-TP (Silvex)	0.05	0.05
1,2,4-Trichlorobenzene	0.07	0.07
1,1,1-Trichloroethane	0.2	0.2
1,1,2-Trichloroethane	0.003	0.005
Trichloroethylene	zero	0.005
Vinyl chloride	zero	0.002
Xylenes (total)	10	10

[a] Each water system must certify, in writing, to the state (using third-party or manufacturer's certification) that when acrylamide and epichlorohydrin are used in drinking water systems, the combination (or product) of dose and monomer level does not exceed the levels specified, as follows: **Acrylamide** 0.05% dosed at 1 mg/L (or equivalent) and **Epichlorohydrin** 0.01% dosed at 20 mg/L (or equivalent).

Table 9.1f U.S. National Primary Drinking Water Regulations
Radionuclides

Radionuclides	**MCLG** (Mg/L)	**MCL** (mg/L) or **TT**
Alpha particles	None	15 picocuries per liter (pCi/L)
Beta particles and Photon emitters	None	4 millirems per year
Radium 226 and Radium 228 (combined)	None	5 picocuries per liter (pCi/L)
Uranium	Zero	30 Ug/L as of 12/08/03

9.2 Profile of U.S. Public Drinking Water Systems

9.2.1 NUMBER OF SYSTEMS

Table 9.2 shows the distribution of U.S. public drinking water systems by size and whether they use surface water or ground water as their water source. Ground water systems are generally smaller than surface water systems, with the median ground water system serving less than 500 people, compared to the median surface water system serving about 3,000 people. Consequently, while 80% of community water *systems* use ground water, those systems serve only 34% of the *total population*. On the other hand, almost half of the U.S. population (44%, totaling over 120 million Americans), is served by systems that provide water, primarily from surface water sources, to 100,000 or more people. Figure 9.1 and 9.2 show that almost 3.4 (72%) of the estimated 190,000 public water systems in the U.S., are *non*community systems, although they serve only a small fraction (under 10%) of the U.S. population (EPA, 1997). The 347 "very large" systems, serving over 100,000 residents each, collectively serve over 112 million people, about 42% of the population served by community water systems.

Unit costs of service are affected by both scale and population concentration, and small and rural systems tend to have dramatically higher costs. Average annual drinking-water costs per household are two to three times higher in systems serving 25-500 people than in systems serving over 100,000 people (Immerman, 1987; Raucher et al., 1994). Also see discussions in Chapter 4 and Chapter 9.

Table 9.2 U.S. Community Drinking Water Systems by Source of Water and Size

System size (Number of people served)	Ground water source		Surface water source	
	Systems number (% of total)	Population millions (% of total)	Systems number (% of total)	Population Millions (% of total)
25-500	28,829 (53%)	4.5 (2%)	3,075 (6%)	0.6 (<1%)
501-3,300	10,414 (19%)	14.1 (6%)	3,626 (7%)	5.7 (2%)
3,301-10,000	2,512 (5%)	14.4 (6%)	1,844 (3%)	11.0 (4%)
10,001-100,000	1,372 (3%)	34.5 (14%)	1,904 (4%)	56.6 (22%)
Over 100,000	68 (<1%)	18.9 (7%)	279 (<1%)	93.6 (37%)
Total	**43,195 (80%)**	**86.4 (34%)**	**10,728 (20%)**	**167.4 (66%)**

Source: US EPA Safe Drinking Water Information System, 2000.

The number of community water systems (CWSs) in the U.S. has not changed much in the past 20 years or so. However, several factors may encourage consolidation in the coming decades, including increased population densities, more stringent regulatory requirements (especially mandatory operator training and qualifications), the search for operational efficiencies, and pressures to coordinate demands on both surface and ground water resources (Levin et al, 2002; Nikas and Katsiamakis, 2002). The U.S. pattern of many decentralized local water systems is also common elsewhere, but an increasing number of industrialized countries are moving to consolidating operations and ownership, with some emphasizing the role of the private sector (CBO 2002; Nikas and Katsiamakis, 2002). For instance, in Great Britain about 20 private regional companies provide almost all wastewater services and most of the drinking water in England and Wales.

9.2.2 SYSTEM OWNERSHIP

Ownership of CWSs can fall into one of several categories. They can be public entities, part of the local government (municipal or county), an independent agency, a special district, or they can be privately owned. While more than half of the CWSs in the U.S. are privately owned, together those systems serve only about 16% of the population (Table 9.3). Private water companies are operated much like public utilities, subject to rate-of-return regulations administered by state public utility commissions.

Table 9.3 U.S. Community Drinking Water Systems by Ownership and Size

System size (Number of people served)	Publicly Owned		Privately owned	
	Systems number (% of total)	Population millions (% of total)	Systems number (% of total)	Population millions (% of total)
25-500	7,353 (14%)	1.8 (<1%)	23,023 (44%)	3.2 (1%)
501-3,300	9,892 (19%)	14.6 (6%)	3,761 (7%)	4.7 (2%)
3,301-10,000	3,671 (7%)	21.5 (9%)	598 (1%)	3.4 (1%)
10,001-00,000	2,804 (5%)	77.3 (31%)	421 (1%)	12.5 (5%)
Over 100,000	284 (<1%)	92.8 (37%)	56 (<1%)	16.8 (7%)
Total*	**24,004 (46%)**	**207.9 (84%)**	**27,859 (54%)**	**40.6 (16%)**

* 2,060 systems serving 5.3 million people had unspecified ownership. Source: see Table 9.2

Publically owned water utilities, if not operating as regulated public utilities, generally must answer to the voters. Recently, a hybrid arrangement is developing of publicly owned water systems that contract with private firms to operate and manage them (CBO 2002; Nikas and Katsiamakis, 2002).

Both drinking water and wastewater systems are natural monopolies, and are almost always provided by a single enterprise. Hence, without the threat of competition, it is unlikely that the social and economic benefits of private ownership of water utilities will be realized. There is both theory and some data to suggest that private systems *can* be more successful in raising investment funds (Haarmeyer, 1992) and charging higher rates for water services (CBO, 1987). This will be increasingly important to continue providing high quality drinking water, especially in view of our deteriorating public water infrastructure. There is recent evidence that the number of independent investor-owned water utilities is declining as more and more are being acquired by other larger utilities, especially European companies, as discussed in Chapter 14 (Nikas and Katsiamakis, 2002).

9.2.3 OPERATING CHARACTERISTICS

The number and percent of drinking water systems that provide treatment has steadily increased throughout the second half of the twentieth century. In 1976, almost 60% of systems serving under 1,000 people provided no treatment whatever; by 1995, that percentage had dropped to about 20% (EPA, 2002). EPA (1997) estimates that total water consumption from CWSs is at least 37 billion gallons (140 million m^3) per day, with 47% going to residential customers and 53% to nonresidential customers. Of water deliveries supplied by CWSs to nonresidential customers within their systems, EPA reported that 43% went to commercial users, 29% to wholesale customers (often other water systems supplying residential customers), 21% to industrial users, 8% to government, and about 1% for agricultural uses.

Treatment characteristics differ dramatically by system size, while requirements are largely determined by water source, ground or surface. In the 1986 Survey of Community Water Systems (Immerman, 1987), over 80% of very large water systems provided conventional drinking water treatment, including coagulation and sedimentation, compared to only 20% of systems serving 500-3,300 people and only 10% of systems serving under 500 people. That disparity remained, and in 1995, about 96% of very large systems provided conventional treatment compared to about 40% of systems serving under 3,300 (EPA, 2002). Similarly, almost 80% of very large systems have professional water treatment managers, compared to about 35% of systems serving under 3,300 (Immerman, 1987). Source water protection efforts were reported by 45% of systems serving over 3,300 compared to 32% of smaller systems reporting such efforts (EPA, 1997).

Drinking water treatment requirements are largely determined by water source, ground or surface. Table 9.4 shows some differences in treatment practices between water systems served by ground versus surface waters and by system size in 1995 (EPA, 2002). Operating costs also differ significantly by system size, with average costs per

1,000 gallons almost doubling from \$131 (\$35/m^3) in systems serving over 100,000 customers to about \$230 (\$61/m^3) in systems serving 1,000 or fewer (see Chapter 4). Examination of a standard financial ratio (operating ratio is obtained by dividing total operating revenues by operating and maintenance expenses) suggests that over 30% of very small systems (serving under 500 people) are, on average, not financially healthy (EPA, 1997; Nikas and Katsiamakis, 2002).

9.3 Challenges Facing U.S. Public Drinking Water Systems

Although most of the 280 million Americans have access to generally safe and reliable drinking water, myriad challenges face the water industry at the turn of the millennium. Solutions to each of these problems may require institutional changes, and will certainly require the best technological and social ingenuity.

9.3.1 INFRASTRUCTURE NEEDS

The major challenge facing U.S. public drinking water systems at the turn of the millennium relates to the *poor state of public water infrastructure* (Levin et al., 2002; Nikas and Katsiamakis, 2002). Numerous studies have shown massive gaps in infrastructure maintenance. In 1994, the Natural Resources Defense Council issued a report, entitled Victorian Water Treatment Enters the 21st Century, which was soundly decried for its alarmist note (NRDC, 94). The NRDC noted the dilapidated condition of much of U.S. public infrastructure, with its attendant public health implications. By 1997, under Congressional mandate EPA issued its own 'Needs Assessments', estimating about \$300 billion in capital needs for drinking water over the next 20 years (EPA, 2002). Shortly thereafter, the Congressional Budget Office issued its own review, projecting annual needs for drinking water of about \$21 billion, which represents a \$9.4 billion annual funding gap (CBO, 2002). Clearly, the needs are enormous and will require concerted public and private efforts to upgrade our deteriorating public water infrastructure, for both drinking and wastewater.

9.3.2 CLIMATE CHANGE

Climate change will also likely challenge the U.S., affecting both the quality and quantity of our water supplies. Surface water will be affected by global warming, with evaporation rates rising due to a warmer atmosphere. The warming of the oceans may lead to sea level rise, with a potential for increased seawater infiltration of coastal aquifers (Levitus et al, 2000; Borrelli, 1988; Levin et al., 2002). Warmer temperatures may also contribute to increased harmful algal blooms, for which there is already global evidence (Harvell et al., 1999; Lagos, 1998), and will provide a more hospitable environment for many pathogenic microorganisms. Warmer temperatures will also result in increased water usage.

Table 9.4 U.S. Drinking Water Treatment by Water Source and System Size

	System Size (people served)				
	500 or under	501-3,300	3,301-50,000	Over 50,000	All Sizes*
Groundwater systems	**Percentage of Systems**				
No treatment	26	13	5	0	
Aeration	15	48	74	87	33
Disinfection/Oxidation	89	95	99	99	92
Iron and Manganese removal	21	45	68	74	34
Flocculation/Coagulation	20	41	72	92	33
Filtration	25	48	82	95	39
Organics removal	14	35	62	77	27
Corrosion control	20	46	77	91	36
Surface water systems					
No treatment	2	0	0	0	
Aeration	1	10	11	59	12
Disinfection/Oxidation	98	100	97	100	99
Iron and Manganese removal	1	10	15	63	14
Flocculation/Coagulation	48	85	99	100	84
Filtration	75	82	93	96	89
Organics removal	3	10	37	79	23
Corrosion control	23	49	70	87	62

* Percentages do not add to 100% since most systems use multiple treatment methods. Source: EPA 1997.

9.3.3 MICROBIAL CONTAMINATION

Myriad studies have shown widespread *microbial contamination* of U.S. drinking water sources by numerous pathogens (Hurst, 1991; Rose et al; 1991; LeChevallier et al., 1991). In addition, recent empirical and epidemiological studies suggest that 10% or more of gastrointestinal disease in the U.S. is related to drinking water (Payment et al., 91; Schwartz et al., 1997; Schwartz et al., 2000). However, the control of waterborne disease through the use of disinfectants, such as chlorine, is complicated by increasing evidence of the carcinogenicity and other *health effects of disinfection by-products* (Morris et al., 1992; EPA 1998; Bove et al., 1995; Waller et al., 1998). Global warming may well create even more hospitable conditions for waterborne pathogens. Finally, consumers and especially the agriculture and aquiculture sectors annually dump tons of antibiotics, medical wastes, and other biologically active substances into surface waters, providing both the impetus and opportunity for waterborne pathogens to mutate and develop resistance to current pesticides and antibiotics (Gaudreau and Gilbert, 1998; Hignite and Azaroff, 1997; Halling-Sorenson et al., 1997; Daughton and Ternes, 1999; Levy, 1998). These issues will certainly challenge both the public health and drinking water communities in the next century.

9.3.4 GROUND AND SURFACE WATER SOURCES

In general, U.S. *surface waters* have greatly benefited from the passage of the Clean Water Act in 1972; nonetheless, virtually none of them are drinkable without treatment. About two-thirds of the U.S. population is served by surface waters. A recent survey of all 50 states conducted by EPA listed problems resulting from anthropogenic activities as the major water quality concerns including siltation, nutrient-loading, pathogen contamination, oxygen-depletion, metals, habitat alteration, and agricultural chemicals (EPA, 2000). Across the U.S., agricultural activities are the most extensive cause of water pollution, affecting 70% of impaired lakes.

Ground water supplies about one-third of the U.S. population. The major problems facing groundwater systems are contamination and excessive extraction (Levin et al., 2002). Residence times in ground water are much longer than in surface water, so "short-term" contamination of soil or surface waters that feed ground water can result in long-term contamination of aquifers. Remediation, where possible, is extremely expensive. And even in aquifers that can be recharged (refilled) by rainfall, current extraction rates exceed the precipitation recharge rates in many basins (EPA, 1995).

9.3.5 URBAN DEVELOPMENT AND LAND USE PRESSURES

Land use pressures will continue to present challenges to U.S. drinking water systems (Levin et al., 2002). Population growth; continued overuse and contamination of groundwater aquifers; fouling of surface waters used for drinking, including especially eutrophication of freshwater bodies; and increasing competing needs for water for recreational and industrial purposes will aggravate existing stresses on our water resources.

References

Baker, M.N.,1949. *The Quest for Pure Water.* The American Water Works Association, New York.

Borrelli, P., 1988. *Crossroads: Environmental Priorities for the Future.* Island Press, Washington, D.C.

Bove, F., M.C. Fulcomer, J.B. Klotz, J. Esmart, E.M. Dufficy, J.E. Savrin, 1995. "Public drinking water contamination and birth outcomes," *Am J Epidemiol* 141: 850-862.

Congressional Budget Office, 1987. *Financing Municipal Water Supply Systems.* US Congressional Budget Office, Washington, D.C.

Congressional Budget Office, 2002. *Future Investment in Drinking Water and Wastewater Infrastructure.* US Congressional Budget Office, Washington, D.C.

Daughton, C.G. and T.A. Ternes, 1999. "Pharmaceuticals and personal care products in the environment: agents of subtle change?" Envir Health Perspec 107 (suppl 6): 807-938.

EPA, 1995. *SAB Report: Safe drinking Water, future trends and challenges.* EPA-SAB-DWC-95-002. Washington, D.C.

EPA, 1997. *1995 Community Water System Survey.* US EPA Office of Water; EPA 815-R-97-001. Washington, D.C. available at http://www.epa.gov/safewater/cwsreprt.pdf

EPA, 1998. *Interim Enhanced Surface Water Treatment Rule.* 63 FR 69477. December, Washington, D.C.

EPA, 2000. *National Water Quality Inventory: 1988 Report to Congress.* EPA 841-S-00-001. Washington, D.C.

EPA, 2002. *The Clean Water and Drinking Water Infrastructure Gap Analysis.* US EPA Office of Water; EPA 816-R-02-020. available at http://www.epa.gov/safewater/gapreport.pdf.

Fair, G.M., and J.C. Geyer, 1958. *Elements of Water Supply and Waste-Water Disposal.* John Wiley and Sons, New York.

Gaudreau, C. and H. Gilbert, 1998. "Antimicrobial resistance of clinical strains of *Campylobacter jejuni* subsp. *jejuni* isolated from 1985 to 1997 in Quebec, Canada," *Antimicrob Agents Chemother* 42: 2106-2108.

Gostin, L.O., Z. Lazzarini, V.S. Neslund and M.T. Osterholm, "2000. Water quality laws and waterborne diseases: *Cryptosporidium* and other emerging pathogens," *Am J Public Health* 90(6): 847-853.

Haarmeyer, D., 1992. *Privatizing Infrastructure: Options for Municipal Water-Supply Systems.* Policy study 151. The Reason Foundation, Los Angeles, California.

Halling-Sorenson, B., S.N. Nielsen, P.F. Lanzky, F. Ingerslev H.C. Holten-Lutzhoft and S.E. Jorgenson, 1998. "Occurrence, fate and effects of pharmaceutical substances in the environment - a review," *Chemosphere* 36(2): 357-393.

Harvell, C.D., K. Kim, J.M. Burkholder, R.R. Colwell, P.R. Epstein, J. Grimes, et al., 1999. "Emerging marine diseases: climate links and anthropogenic factors," *Science* 285: 1505-1510.

Hignite, C., and D.L. Azarnoff, 1997. "Drugs and drug metabolites as environmental contaminants: chlorophenoxyisobuyrate and salicyclic acid in sewage water effluent." *Life Sci* 20(2): 337-341.

Hoxie, N.J., J.P. Davis, J.M. Vergeront, R.D. Nashold and K.A. Blair, 1997. "Cryptosporidiosis-associated mortality following a massive waterborne outbreak in Milwaukee, Wisconsin." *Am J Public Health* 87(12): 2032-2035.

Hurst, C.J., 1991. "Presence of enteric viruses in freshwater and their removal by the conventional water treatment." *Bull World Health Org* 69(1): 113-119.

Immerman FW., 1987. *Final Descriptive Summary: 1986 Survey of Community Water Systems.* Prepared for: US EPA Office of Drinking Water; Washington, DC; by Research Triangle Institute; Research Triangle Park, North Carolina (July).

Karalekas P.C, J.R.Trax, 1989. "Compliance by public water supply systems with National Primary Drinking Water Regulations," In: *Safe Drinking Water Act: Amendments, Regulations and Standards* (Calabrese E.J., C.E. Gilbert and H. Pastides, eds); Chelsea, Michigan; Lewis Publishers: 105-112.

Lagos N., 1998. 'Microalgal blooms: a global issue with negative impacts in Chile." *Biol Res* 31(4): 375-386.

Larson C., 1989. "Historical development of the National Primary Drinking Water Regulations." In: *Safe Drinking Water Act: Amendments, Regulations and Standards* (Calabrese E.J, Gilbert C.E, Pastides H, eds); Chelsea, Michigan; Lewis Publishers: 3-16.

LeChevallier, M.W, W.D. Norton and R.G. Lee, 1991. "Occurrence of *Giardia* and *Cryptosporidium* spp. in surface water supplies." *Appl Environ Microbiol* 57(9): 2610-2616.

Levin R, P.R. Epstein, T.E. Ford, W. Harrington, E. Olson and E. Reichard, 2002. "US drinking water challenges in the twenty-first century." *Environ Health Perspect* 110(supplement1): 43-52.

Levitus, S., J.I. Antonov, T.P. Boyer and C. Stephens, 1998. "Warming of the world's oceans." *Science* 287: 2225-2229.

Levy, S.B., 1998. "Multidrug resistance--a sign of the times." *N Engl J Med* 338: 1376-1378.

MacKenzie, W.R, N.J. Hoxie, M.E. Proctor, M.S. Gradus, K.A. Blair, D.E. Peterson, J.J. Kazmierczak, D.G. Addiss, K.R. Fox, J.B. Rose and J.P. Davis, 1994. "A massive outbreak in Milwaukee of *Cryptosporidium* infection transmitted through the public water supply." *New Eng J Med* 331: 161-167.

McDermott, H. 1973. "Federal drinking water standards -- past, present and future." *J Environ Eng Div ASCE* 99(EE4): 469-478.

Morris, R.D., A. Ardet, I.F. Angelillo, T.C. Chalmers and F. Mosteller, 1992. "Chloramination, chlorination by-products and cancer: a meta-analysis." *Am J Public Health* 82: 955-963.

Natural Resources Defense Council, 1994. *Victorian Water Treatment Enters the 21st Century*. Washington, D.C. (March).

Payment, P., L. Richardson, J. Siemiatycki, R. Dewar, M. Edwardes and E. Franco, 1991. "A randomized trial to evaluate the risk of gastrointestinal disease due to the consumption of drinking water meeting current microbiological standards." *Am J Public Health*; 81: 703-708.

Raucher RS, A.M. Dixon, E. Trabka and J.A. Drago, 1994. "Cost effectiveness of SDWA regulations." *J Am Water Works Assoc* 88 (8): 28-36.

Rose, J.B., C.P. Gerba and W. Jacobowski, 1991. "Survey of potable water supplies for *Cryptosporidium* and *Giardia*." *Environ Sci Tech* 25(8): 1393-1400.

Schwartz, J., R. Levin and K. Hodge, 1997. "Drinking water turbidity and pediatric hospital use for gastrointestinal illness in Philadelphia." *Epidemiol* 8: 615-620.

Schwartz, J., R. Levin and R. Goldstein, 2000. "Drinking water turbidity and gastrointestinal illness in Philadelphia's elderly." J Epidemiol Comm Health; 54: 45-51.

Taylor, F.B., 1977. "Drinking Water Standards Principles and History 1914 to 1976." *J New England Water Works Accoc.* 91(4): 237-259.

US Centers for Disease Control and Prevention, 1999. "Ten great public health achievements -- United States, 1900-1999." *Morbid Mortal Wkly Rpt* (Centers for Disease Control and Prevention) 48(12):241-243.

Waller, K., S.H. Swan, G. DeLorenze and B. Hopkins, 1998. "Trihalomethanes in drinking water and spontaneous abortion." *Epidemiol* 9: 134-140.

Chapter 10

Water System Organization and Financial Decision Making

R. Bruce Billings
University of Arizona

10.1 Monopoly Market is Most Likely

Water distribution is a natural monopoly, in which a single entity can supply each geographic service area at a much lower cost that any group of two or more competitive water providers. Dual water systems exist but are generally owned by the same utility in order to deliver different qualities of water. A potable water delivery system may be augmented by a reclaimed water delivery system for industrial or commercial customers.

10.2 Private vs. Public Provision of Water

10.2.1 GOALS OF PUBLIC AND PRIVATE PROVIDERS

There is no clear answer to the question: "Are water customer's served better by a government water utility or a private but government regulated firm?" Publicly owned water utilities take a number of specific forms, including municipal departments, government owned and controlled enterprises, and independent water supply districts (often irrigation districts) with their own elected or appointed boards of directors. While some of these public utilities have single purpose elected boards, many others are

D.E. Agthe et al. (eds.), Managing Urban Water Supply, 147–165.

overseen by elected officials for whom the water utility is only one of many responsibilities.

Most public water utilities say that their goal is to provide safe reliable supplies of water at the lowest possible cost. Keeping rates low is often a prime goal, since customers are voters. Timmons (2002) shows that 80 percent of municipally owned water utilities in California price their water at less than marginal cost. His research also found a significant negative correlation between the variance of the income distribution in a community and the ratio of marginal price to average variable cost of water production, indicating that municipalities with wider income distributions have lower percentages of total cost covered by water related revenue. This suggested that water rates are kept low to help offset the regressive nature of the fixed charges and to benefit low-income consumers.

Private water utilities' primary goal is to make a profit for their stockholders by providing safe reliable supplies of water at whatever prices are allowed by the government regulators. Prices must be set in recognition of the demand for water so that the utility can obtain sufficient revenue to operate, attract stockholders and sell bonds.

The desire to increase profits by increasing efficiency may lead a private utility to operate at a lower cost than a publicly operated utility. However, this drive toward efficiency is blunted if utility regulators set rates to guarantee a fixed rate of return on invested capital. In this case, higher costs of operation are simply passed on in higher water rates. Unfortunately, the same phenomena occur when the utility is owned and operated by the public sector. Managers have a great incentive not to "rock the boat" and a very limited incentive to operate efficiently. When employees are hired under civil service status or union contract, it is much easier to shift ineffective employees to some innocuous job than to dismiss them.

The Congressional Budget Office (1987) reported that the prices charged by private utilities in the U.S. exceed those imposed by public utilities by 30-80%. Prices are generally lower among government run utilities because they are more likely to defer expensive maintenance projects and do not include depreciation of capital equipment or a return to invested capital as costs. They are also more likely to be subsidized through other governmental revenues. Public utilities sometimes finance capital expenditures through bonds repaid by property taxes instead of utility revenue, thus shifting part of water costs from users to general taxpayers.

Equity is a concern to both government providers and government regulators of private providers when commodity charges, fixed charges, and connection fees are set. Household water use and therefore expenditures on water commonly increase as household income increases but decrease as a percentage of household income. This phenomenon has been observed by Timmins (2002) in California, Whittington (et al., 1991) in Nigeria, and Israel (2001) in Bolivia. Fixed periodic charges and connection fees tend to be regressive since they are a higher percentage of income for lower income households. In many developing nations, high connection fees act as a barrier to entry for low-income households. These households then use water from nearby untreated sources or purchase water from mobile water vendors who tend to sell their water for a higher per unit price than that of piped water. It is not uncommon for several households to be served by a single water tap. Israel (2001) suggests that households without piped water would experience a welfare gain if they were included in the piped water system. If the welfare gains of extending the piped water system are large enough, there is a strong argument for government subsidized piped water.

10.2.2 FUNDING DIFFERENCES

Final customers and/or taxpayers ultimately pay for water supply regardless of how its provision is organized. An advantage of privately owned and operated utilities is the ability to obtain capital to finance construction and improvements through the issuance of stock and bonds. Of course, the cost and availability of these funds depends on the existence of a predictable and adequate demand for water which will generate a revenue stream sufficient to pay all operating costs, including interest and dividends to stockholders. In contrast, a public utility can often raise funds by selling government backed bonds to be repaid with tax revenue if utility income falls short. If the governmental entity has a good record of repaying bond interest and principle when due, governmental bonds may carry a lower interest rate than bonds issued by a private utility firm, giving the public entity organization form a cost advantage. In the United States, the public utility may also have the advantage of issuing bonds with interest payments exempt from the federal (and sometimes state) income tax, thus yielding a lower than market interest rate. Raising water rates to pay for needed capital outlays may be easier for a private utility facing a rate commission hearing than for a public utility which is constrained by politicians facing reelection. Water rate increases proposed by the public water utility in Tucson, Arizona, receive far more publicity and public discussion than proposed natural gas or electricity rate increases, subject to Arizona Corporation Commission approval, even though these energy bills far exceed water bills for most households (Martin, et. al. 1985). In this case, the elected regulators who approve rates for the privately operated utilities are further removed from the rate payer-voters than the city council which sets water rates for the city owned utility. However, when water utilities were privatized in Bolivia and water rates were subsequently increased dramatically to finance capital outlays, customer protests forced the government to roll back the rate increases and rescind privatization.

10.2.3 RECENT U.S. AND U.K. EXPERIENCE

Several recent studies in the U.S have found widespread under-investment in both drinking water and wastewater systems. The estimates of additional needed investment in water supply range from $7 billion to $15 billion per year in addition to a range from $12 billion to $19 billion for wastewater facilities (EPA, 2001; Water Information Network, 2000; American Society of Civil Engineers, 2001). Taking the midpoints of these ranges suggests a needed increase in utility income and spending of $26 billion per year. These increases compare with estimated annual total water utility spending of $36 billion and wastewater utility spending of $25 billion. To fund these needed capital outlays would require an increase of 43% from current spending of $61 billion per year. With a typical price elasticity of demand of -0.5 and corresponding price elasticity of revenue of +0.5, this would require an increase in combined water and sewer rates of about 86%! Since most water bills in the U.S. absorb less than one percent of household income, such increases would not be unimaginable or unduly burdensome. Nevertheless, the two non-governmental groups preparing these cost estimates advocate massive federal government subsidies. Such subsidies, however, are inefficient since they continue the tradition of pricing water below its cost of provision with the resulting over use of this resource.

The degree of under investment in water utilities is also suggested by the impact of privatization in the United Kingdom. In the two years following privatization in 1989, investment spending increased by more than 80% (Haarmeyer, 1992). There are about 54,000 water utilities in the U.S. compared to less than 30 in the U.K. (US EPA, 2001). The large number of suppliers in the U.S. is partially the result of the size and population density of the country. However, it also results from an historic preference for local control of water utility services, the ease of entry into the industry which has prevailed until recent years, and the historic unwillingness of utility commissions to allow adequate water rates to make private investment attractive. In the western U.S., many subdivision developers have in the past found it profitable to start their own water utility as part of the land development process. With increasingly stringent environmental and health regulations imposed on utility companies, many of these factors are no longer in place. However, local governments are often reluctant to merge utility operations into more efficiently sized units for political reasons. Were widespread privatization of water utilities to occur in the U.S. as it has in the U.K., consolidation would undoubtedly follow. Some consolidation among private water utilities has been occurring in the U.S., and this trend is expected to continue. These is also a tendency for larger municipal utilities to buy out small private utility operations in their vicinity, either by invoking eminent domain or by voluntary sale in response to dwindling availability of raw water and increasingly stringent water testing and treatment requirements.

10.3 Taxes, Government Spending and Water Utilities

Many governmental entities impose taxes on water utilities, including franchise taxes for the right to provide water service, sales taxes imposed on water bills, and in the case of private providers, income taxes on net income. In many areas, significant funding for other governmental programs is obtained by these taxes. The government owned utility may also be operated in such a way as to generate substantial surpluses (profits) which are used to support other governmental programs.

Conversely, many jurisdictions provide water service below cost. Capital expenditures may be financed with bonds repaid out of property or other tax sources while only the operating cost of the utility is funded from water rates. In some cases, limited quantities of water are provided at a zero explicit price, with the entire cost paid from tax revenue.

While government subsidies could be provided to private water utilities, it is politically easier to subsidize a governmental water utility. Within a governmental entity, both costs and revenues are somewhat fungible, so items that are properly attributable to the water system can be entered under other categories, such as roads. For example, when paving a road and installing a water main under it at the same time, costs can be shifted between these two projects in either direction. Likewise, creative accounting can shift costs from other areas to the water department.

10.4 Influencing Urban Development

Regulating urban growth and development via water supply policies can be a useful policy option for government water providers. Governments can refuse to invest in water systems to support development of housing, commercial enterprises, and industrial facilities in areas where development is not considered desirable or to prevent urban sprawl. They could also implement measures to limit development to protect a watershed or environmentally sensitive areas.

For efficient growth planning, urban planners must determine where new development should take place and provide water supply and other governmental services such as roads only to those chosen areas. The government agency that owns the water utility simply enforces bans on new connections and well drilling in the undesirable area as well as a requirement that all new water users in the restricted area must have a city water connection to obtain a building permit. Sometimes, the ban on new connections is only temporary until the city chooses to develop a specific area by improving the water lines. In other cases, the agency might consider offering tradable permits (called transferable development rights) to compensate those unable to develop their land. Purchase of these permits from landowners in the area of banned development would be required to obtain water connections for development in the allowable area. Thus, the landowners in the banned area would be compensated for their inability to develop their sites.

Water providers can also consolidate their holdings to better control the water supply and environmental damage in their service areas. A merger of providers pumping out of the same aquifer is likely to result in better aquifer management. In areas of falling water tables, increasing costs of individual wells and the drying up of wells as the water table falls may force individual well owners to abandon their wells to join a better managed local or regional water system.

High water prices, especially to outlying areas, may also be used as a way to discourage further extensive urban development. However, if people are allowed to occupy urban fringe areas without water supplies, such policies as a means of controlling urbanization of formerly rural land areas may be useless or even counter productive. Having large numbers of squatters, for example, in areas without water or sewer services tends to create significant public health problems from unsafe water supplies and untreated sewage.

10.5 Public Health Issues

It is often supposed that a government run utility will be more concerned with water quality and with providing exemplary service than an equivalent private enterprise. However, this is not necessarily the case. In the United States, before the federal government became heavily involved in water quality policies, some of the worst water polluters were city governments that dumped raw sewage into rivers, lakes or the ocean. City run utilities have also been just as slow as privately run utilities to respond to new water testing and treatment requirements imposed by the federal government. In both cases, utility management has been reluctant to change operating procedures or to adopt new more expensive water testing and treatment programs.

More extensive water treatment typically requires higher expenditures that are reflected in higher water rates. Administrators are often reluctant to raise rates, even when the marginal benefit of higher quality potable water is much greater than the marginal cost. A related health issue concerns the collection, treatment, and disposal of wastewater. Since wastewater is created by the indoor use of water from the utility, its treatment and disposal cost should be included in water bills.

10.6 Benefit-Cost and Project Analysis[27]

Water utilities are often faced with the question of whether a proposed project is worthwhile, or which of several alternative projects should be undertaken. For example, an analysis of existing water mains may reveal numerous leaks. Is the cost of main replacement justified by the value of the water saved and the avoidance of collateral damage caused by leakage? Or a city may have several alternative projects available to address a water supply problem, including a new dam on a river, new deep aquifer wells, or end use conservation programs to reduce demand.

For example, New York City recently decided to invest in upgraded protection of its Catskills watershed to avoid the high cost of building a filtration plant. The estimated cost of building a suitable filtration plant was $6 to $8 billion, with annual operating costs of another $300 million. The City estimated that with an expenditure of less than $1.5 billion they could maintain the quality of their water supply. The funds would be spent for land acquisition, new watershed rules and regulations, and financial assistance to local communities to boost local economies and promote environmental quality. The decision by New York to invest in the protection of its drinking water source was supported by a scientific review by the US National Research Council (2000). Careful analysis, including the following steps, is required to make good choices.

[27] There are a number of good books on project evaluation, including Mishan (1973) and Bruce (1998).

10.6.1 DEFINE THE OBJECTIVES OR GOALS

The objective to be attained or the problem to be solved should be carefully defined. The more specific the goals are, the easier it is to analyze the potential effectiveness of alternative proposals. Carefully defined goals, however, may make reaching consensus more difficult, and are often opposed by decision makers who do not want to be held accountable for results.

10.6.2 USE BRAINSTORMING

Brainstorming is often an effective way to find and clarify alternative ways of achieving the desired results. Then list all of the potential costs and benefits of each alternative. Encourage participation in this process, and allow consideration of a wide range of options. There are often several ways in which some goal could be achieved, and open discussion may bring forth useful new ideas. For example, where untreated black water is a public health problem, it could be dealt with by installing a sewage collection and treatment system or by supplying each household with a composting toilet.

10.6.3 QUANTIFY EACH COST AND BENEFIT

Each cost and benefit identified should be quantified using current monetary values. Future costs and benefits should *not* be adjusted for expected inflation. While some items have obvious monetary values, others do not. Creativity is required to assign values to some of the costs and benefits. For example, one cost of damming a river and withdrawing water is reduced flow, which may make the river downstream less viable for fishing, shipping, and other activities. These costs should not be ignored, even though exact quantification is not possible. Similarly, side benefits such as better flood control should be estimated and included. However, double counting should be carefully avoided. The wages paid for labor to build the project is a cost, and the income taxes paid because of those wages is a benefit (for a government project). The income earned by these laborers should not be counted as a benefit of the project, even though it is a benefit to the local economy. If currently unemployed resources are to be used by the project, their cost may be estimated as thc opportunity cost of putting them to work. This may be significantly lower than their monetary cost.

10.6.4 ADJUST FOR UNCERTAINTY

The true value of some of the costs and benefits can be known with certainty only after the project is completed. Thus an adjustment for uncertainty may be useful. For example, if there is a 60% chance of receiving a benefit with an estimated value of \$50,000, a 25% chance that it will be worth \$20,000, and a 15% chance of it having no value, the estimated benefit should be computed as: .6 x \$50,000 + .25 x 20,000 + 0 = \$35,000. In a more complex setting, a detailed sensitivity analysis could be undertaken by estimating values over the possible range of assumptions.

10.6.5 DISCOUNT FUTURE VALUES

First, place each cost and benefit on a time line. For example, if the proposed project is a new dam, most of the cost will occur in the first or second year and the benefits will be received over many years. Then discount all of the costs and benefits back to the beginning of the project in order to obtain the present estimated value of benefits and costs or net present value (NPV). The discount rate should reflect the social opportunity cost of capital. When borrowing local or internal funds, the interest rate for the loan is the appropriate discount rate, since the project should provide sufficient benefits to repay the interest and principle. When all future benefits and costs are measured at current price levels (not adjusted for future expected inflation) the interest rate used in these computations should also be inflation free. Thus if funds are borrowed internally the appropriate interest rate is the borrowing rate, often referred to as a nominal rate, *minus* the current inflation rate, yielding what is know as the *real interest rate.* For example, if the current interest rate is 9.3 percent and the inflation rate is 4.2 percent, the real interest rate, which should be used for discounting, is 5.1 percent. This should still provide sufficient value to repay the loan, since the value of services provided by the project will also tend to increase in nominal value in step with inflation. For projects with benefits accruing over very long time periods, the interest rate used for discounting may be even smaller. Weitzman (1998) argues for a discount rate that declines with time for projects such as watershed enhancement through control of pollution and aquifer enhancement through reduced over-pumping because the expected benefits extent into the indefinite future. With standard discount rates, such projects might never be undertaken since net benefits often appear to be negative. If capital must be borrowed from abroad the choice of an appropriate discount rate is more difficult. In general, the project would need to generate benefits sufficient to facilitate loan repayment. Thus the discount rate would be the loan rate. However, this is complicated by unexpected changes in the exchange rate when the loan is in a different currency. The project managers must be certain that the debt service can be met, and consider the effects on the balance of payments.

10.6.6 COMPARE DISCOUNTED BENEFITS AND COSTS

The comparison of benefits and costs is often accomplished by taking the benefit-cost ratio. If the ratio is greater than unity, the project would probably be accepted if it is mutually exclusive from other projects and credit rationing does not exist. A net present value constructed by subtracting the present value of cost from the present value of benefits could also be used. If the net present value is positive, the project should be accepted if it is mutually exclusive from other projects and credit rationing does not exist. When there are sufficient resources to undertake all proposed projects, either the benefit cost ratio rule or the net present value rule is sufficient as a decision rule. When resources are unlimited, all of the investments that have a positive net present value should be undertaken. However, in the practical world there are usually limits on available investment funds. Public water districts may be limited to amount of money raised by bond sales, a preference by voters to limit debt, or by a desire to keep their debt/revenue ratio low to protect their credit ratings or borrowing capacity. Private water companies may wish to keep their debt/equity ratios low to satisfy stockholders, maintain good credit ratings, and keep their interest payments low. When resources are limited and only a few of many worthwhile projects can be undertaken, they should be selected to maximize the total net benefits of the projects to be undertaken. Under conditions of limited resources, the proper decision rule is to compare Net Present Values for each of the proposed projects. The net present value of all possible combinations of projects must be considered to determine the net present value of that combination of projects which yields the highest combined net present value for the limited invested funds (Neveu, 1989). However, under no circumstances should an investment with a zero or negative net present value be included in the selected combination. The net present value approach provides an appropriate way to compare several mutually exclusive projects. The project with the largest net benefit should usually be selected, although if one project has a dramatically larger benefit cost ratio than the others it might be preferred. For example, a large project with discounted benefits of $70 million and costs of $50 million for a net benefit of $20 million might be passed over for a smaller project with benefits of $25 million and costs of $12 million.

10.7 Considerations for Capital Repayment by General Taxation

When project costs are to be repaid by general taxation, the economic inefficiency resting from taxation must also be considered as a cost. That is, to raise a dollar in tax revenue imposes costs on individual taxpayers and government beyond the revenue gained. Government must spend resources collecting and enforcing the tax, the taxpayer must spend resources complying which includes filling out tax forms, keeping records in the form specified by government, etc. and may also suffer reduced profits or utility because of tax induced changes in behavior. Tax avoidance and evasion schemes are well known, and result in their own costs in lowered productivity. Thus to raise an additional dollar in tax revenue may impose total costs of society ranging from slightly over a dollar to a dollar and a half or more. These indirect costs of financing must be included in the overall cost calculation to avoid understating total project costs.

10.8 Benefit Cost Analysis: An Example

The following analysis of the potential use by Tucson, Arizona of water from the Colorado River which is delivered via the Central Arizona Project (CAP) illustrates many of the steps involved in a benefit-cost analysis A major part of this analysis is the estimation of the benefits to the customers of the water utility of using this new water resource. The overall analysis involved estimating a demand function for water, evaluating the safe yield of the aquifer supplying groundwater, determining the incremental cost of the new water supply, discounting, evaluating uncertainty and deriving a benefit cost comparison.[28]

10.8.1 ESTIMATING A DEMAND FUNCTION FOR URBAN WATER

The demand model was estimated in both linear and multiplicative (double logarithmic) form because of uncertainty over the true shape of the demand function. The model was estimated using a pooled time series-cross section sample with 84 monthly observations on household water use from January 1974 through December 1980 for each of 12 demographic districts served by Tucson Water. The demand model, with the results shown in Table 10.1 is:

$$Q = f\,(P, I, H, G, A_1, A_2, T, HT, R, S) \qquad (10.1)$$

Where: Q = Water use/household (100 ft^3/month)
P = Average price of water (1985 dollars)
I = Income/household/month (1985 dollars)
H = Household size (persons)
G = Ratio of households served each month to those in January 1974
A_1 = Percent of heads of household age 55 to 64
A_2 = Percent of heads of household age 65 and over
T = Mean monthly temperature (F)
HT = High temperature (degrees above 58 F)
R = Monthly summer rainfall (inches)
S = Proportion of annual daylight hours in month

Both the linear and multiplicative econometric versions of the demand model were estimated in the dynamic Koyck form, which includes the lagged value of the dependent variable as an independent variable and provides both short-run and long-run estimates of the parameters. The long-run values are shown in Table 10.1 and used in the analysis. Both versions are highly successful in explaining variations in water use. Similar demand models are fully discussed in Billings and Day (1989).

Average price is used in these models because most Tucson customers were found to be unaware of the increasing block rates that had been in effect for over six years (Agthe et al., 1988). Because of the relatively large monthly availability of service charge, most customers faced an average price of water that decreased with increasing use levels. Water rates changed frequently during the study period and the steps in the increasing block rate water tariffs were too small to have a dramatic effect on customer decision making.

---

[28] This analysis is excerpted, from Billings, R.B., 1990. "A Demand Based Benefit-Cost Model of Participation in a Water Project," *Journal of Water Resources .Planning and Management*, Volume 116 (5): 593-609, Copyright 1990 by The American Society of Civil Engineers (ASCE). It appears here by permission of ASCE.

Table 10.1 Demand Model Results

	Linear Model			Multiplicative Model	
Variable	Coefficient	Elasticity	t statistic	Elasticity	t-statistic
P	-0.275	-0.717	17.7	-0.565	20.1
I	0.0078	0.300	11.2	0.296	13.6
H	0.00710	0.136	2.3	0.221	5.0
G	-2.515	-0.190	3.5	-0.110	2.6
A_1	-0.00907	-0.079	2.7	-0.0661	3.8
A_2	0.00999	0.127	3.3	0.171	5.9
T	0.121	0.516	2.2	0.742	8.2
HT	0.156	0.118	2.7	0.0174	8.8
R	-1.303	-0.053	7.8	-0.019	10.3
S	1.462	0.760	6.8	0.980	10.6
Q lagged 1 month	0.308	-	12.3	0.355	18.0
Adjusted R^2	0.839			0.888	
F	479			783	
Rho	--			-0.10	
D-W	2.019			2.022	
Durbin's H	0.484			0.433	

Notes: Long run parameters shown. R^2 shows explanatory power in terms of Q, not log Q.

Water use and prices were determined from utility records while demographic characteristics for each district were obtained from survey data. Smoothing techniques were used to convert the periodic demographic observations into monthly data. While the income variable primarily shows differences among districts it also includes variations over time. The demographic variables primarily serve to control for differences among districts, so that the income and household variables show long run adjustments to these factors. All monetary values were adjusted to constant 1985 dollars using the national consumer price index. Average temperature and summer rainfall were obtained from the U.S. weather bureau. The high temperature variable was added to account for intensified yard irrigation demand during hot summer months.

10.8.2 USING THE DEMAND MODEL TO ESTIMATE BENEFITS

Demand functions show the price that buyers would be willing to pay for each possible quantity. With voluntary purchases of water, the price at each point on the demand curve therefore shows the marginal value of that particular quantity of water. The total value of water purchased can be obtained by measuring the area under the demand curve up to the quantity taken. For example, in the demand curve of Figure 10.1, if a consumer buys five units of water at \$3.00 per unit, the marginal or last unit of water purchased must have a value to the consumer of \$3.00. If its use value were less, the consumer would not buy the 5th unit, and if its value were higher the consumer would purchase additional water. Given the demand curve, the total value (not cost) to the consumer of this water is the area under the demand curve from Q=0 to Q=5. In this example, that value is $18.75. If the price of water were to drop to $1.50 and use rose to 10 units, as predicted by the demand curve, the value of the *additional* water to the consumer would be the area under the demand curve between Q=5 and Q=10, which is $11.25. With a price of $1.50 per unit, the additional five units of water would cost our consumer $7.50, yielding a net benefit of $3.75.

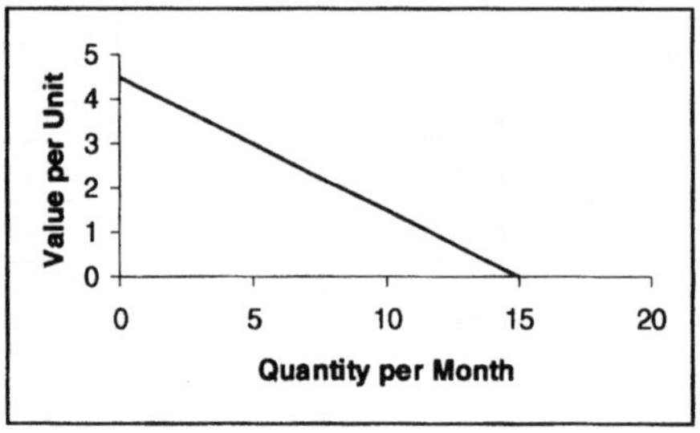

Figure 10.1 Marginal Value of Each Unit of Water

The demand curves estimated from equation 10.1 are shown in Figure 10.2. The estimated demand curves are quite accurate for predicting changes in quantity within the range of data used in estimating the model. However, when it is necessary to estimate changes outside of this range of historic experience, the shape of the demand curve is unknown and must be assumed. Thus the model was estimated in both linear and multiplicative form. In the latter case, the curve has a constant elasticity while in the former it has a constant slope. The linear curve exhibits the property that as the quantity of water falls, its value increases in direct proportion to the drop in quantity. In the multiplicative model, however, as the quantity of water falls its value increases exponentially. At quantities of water which would be available to Tucson consumers under safe yield without imported water, the divergence in value becomes quite large. The linear and multiplicative models provide a range within which the true demand curve is likely to fall.

In applying the demand model to estimate benefits of the CAP, it is assumed that the estimated parameters for all other utility customers are the same as those estimated for single family residential users. This may overstate the overall elasticity of demand, since water use in apartments and commercial establishments may be less price-sensitive than for individually metered homes. However, any overstatement is likely to be small because a substantial proportion of non-residential water, like residential water, is used for irrigation of decorative plantings and play areas. Many large non-residential users have their own private wells and are therefore excluded form this analysis. Thus variations in price and other variables are likely to have similar impacts on water use by all types of users.

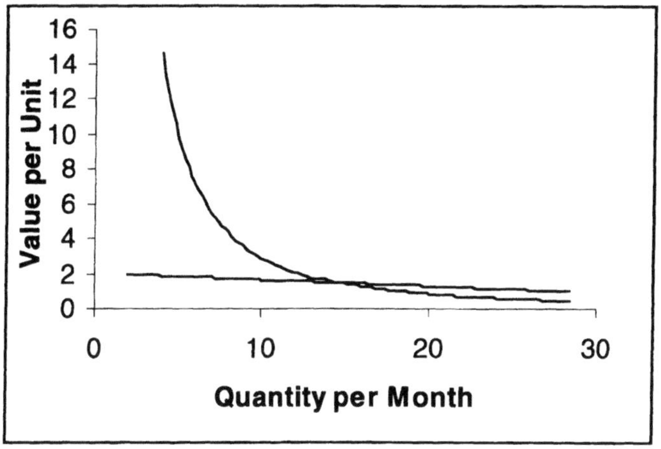

Figure 10.2 Estimated Marginal Value of Water

The benefits models were estimated for the value of incremental project water for a typical consumer. The individual household benefits were then grossed up for all households and all water users. The estimated benefits of CAP water to Tucson Water customers for selected years are presented in Table 10.2, assuming that Tucson Water reduces groundwater pumpage to its share of estimated safe yield during the first full year of CAP operation. Area population is assumed to continue growing at its historic 2.59% per year rate. Additional population growth after 2030 is assumed to receive water from some other new source. Personal income is assumed to continue growing at its historic 2% per year rate in constant dollars while average household size declines from 3.3 persons per household in 1977 to 2.36 in 1991 and 2.2 by 2000, after which it will remain constant. The importance of rising household income is that water use is known to increase with income. With the income elasticity of demand shown in Table 10.1, water demand increases by about 3% for each 10% increase in income. Thus rising incomes increase the value of any given water supply. Continued population growth has the same effect, spreading water supplies over a larger population.

With these assumptions, a typical household in 2000 has 13.2 Ccf (hundred cubic feet) or 37.4 m^3 of water available per month with the CAP but only 4.4 Ccf (12.5m^3) without it, as shown in Table 10.2. The benefit of the additional water from the CAP to a typical household is estimated at $40.43 per month with the exponential model but only $14.99 under the linear version. The two estimates differ since the quantity of water without CAP is much smaller than the current average household use of about 13 Ccf (37m^3) per month and the two demand curves shown in Figure 10.2 have very different values at the 4.4 Ccf (12.5 m^3) quantity.

Table 10.2 Tucson Water Supply per Household

Year	CAP Supply 1000 AF/Year	Q/Household Ccf per Month	
		No CAP	With CAP
1991	60.0	4.7	12.8
2000	83.3	4.4	13.2
2010	109.2	3.9	12.9
2020	135.1	3.3	11.9
2030 & after	148.0	2.7	10.0

10.8.3 DISCOUNTED BENEFITS AND SENSITIVITY ANALYSIS

Benefits were discounted to 1985 dollar values at the 8.626% which is the rate currently used by the federal government to analyze new projects and at 3.25% which is the rate used in the original benefit-cost analysis. This range provides a good indication of the sensitivity of results to the choice of discount rate. Several model options were examined in addition to the standard assumption of the immediate imposition of safe yield and an end to growth of population served by project water when availability drops to 10 Ccf (28 m^3) per household per month.

If safe yield is phased in over 20 years instead of being imposed upon completion of the CAP, the benefits from CAP water are reduced by over 31% under the exponential model at the 8.625% discount rate, but by less than 10% under each of the other models, as shown in Table 10.4. If population served by CAP water continues to grow after 2030, the value of this water increases by 136% with the 3.25% discount rate and the exponential model. With the higher 8.625 rate, discounted benefits increase by less than 15% since the added benefits are so far in the future. Under the linear model, population growth after 2030 has little impact on discounted benefits.

The sensitivity of results to the value of the price coefficient in the demand model was also examined. Benefits under the linear model varied by 4% or less for a 10% change in this parameter. In contrast, benefits under the exponential model increased by nearly 28% for a 10% reduction in the elasticity of demand at the 3.25% discount rate while falling almost 18% for a 10% increase in elasticity. The corresponding changes with the 8.625% discount rate were 21% and 14%.

10.8.4 PROJECT COST

The cost of using CAP water include payments for the water, the cost of constructing and operating a water treatment plant, and new transmission lines to carry the water from the canal and treatment plant into the existing distribution system. Offsetting reductions in cost accrue from reduced pumping costs, possible reduced future expenditures on other remote water sources, and avoided cost of groundwater depletion and subsidence. It is not possible to determine the exact incremental cost of using the CAP because planning for the past ten years has been directed toward getting ready for its use.

Table 10.3 Benefits of CAP for Tucson Water Customers: Exponential Model

Year	Exponential Model Benefits (1985 dollars)			
	Marginal Benefit Per Ccf		CAP benefit/ household/ month	Total Benefit per Year
	Without CAP	With CAP		
1991	10.51	1.74	34.47	9,165,000
2000	12.36	1.77	40.43	13,863,000
2010	16.87	2.06	51.54	22,821,000
2020	25.61	2.64	68.63	39,244,000
2030	39.63	3.97	88.68	65,490,000
2040	40.53	4.37	93.13	69,663,000
2050	42.00	4.83	98.55	74,667,000
2060	44.01	5.35	104.96	80,551,000
2070	46.55	5.95	112.42	87,388,000
2080	49.63	6.62	121.00	95,271,000
2090	53.30	7.37	130.79	104,311,000

Table 10.4 Benefits of CAP for Tucson Water Customers: Linear Model

Year	Linear Model Benefits (1985 dollars)			
	Marginal Benefit Per Ccf		CAP benefit /household/ month	Total Benefit per Year
	Without CAP	With CAP		
1991	1.75	1.45	13.09	3,480,000
2000	1.86	1.54	14.99	5,142,000
2010	2.02	1.69	16.62	7,385,000
2020	2.21	1.90	17.65	10,091,000
2030	2.44	2.18	16.82	12,418,000
2040	2.69	2.43	18.41	13,773,000
2050	3.00	2.74	20.36	15,430,000
2060	3.38	3.12	22.74	17,455,000
2070	3.84	3.58	25.64	19,929,000
2080	4.40	4.15	29.15	22,949,000
2090	5.08	4.83	33.39	26,633,000

Table 10.5 Sensitivity Analysis
Discounted Benefits for First 20 Years and Entire 100 Years
($1,000)

Standard Assumptions				
	Exponential Model		Linear Model	
Discount Rate	8.625%	3.25%	8.625%	3.25%
1991-2010	80,653	174,023	29,607	63,070
All Years	143,652	910,791	44,828	228,788
Population Served by CAP Continues Growing after 2030				
1991-2010	80,653	174,023	29,607	63,070
All Years	164,687	2,145,195	44,910	231,690
Ratio to Standard	1.15	2.36	1.00	1.01
Twenty Year Phase in of Safe Yield				
1991-2010	36,123	86,097	26,502,	57,387
All Years	98,880	821,962	41,717	223,084
Ratio to Standard	0.69	0.90	0.93	0.98
Price Elasticity of Demand and Slope of Demand Curve Reduced Ten Percent				
1991-2010	94,109	203,751	30,325	64,479
All Years	174,150	1,164,474	46,077	238,756
Ratio to Standard	1.21	1.28	1.03	1.04
Price Elasticity of Demand and Slope of Demand Curve Increased Ten Percent				
1991-2010	71,291	153,405	29,094	61,917
All Years	123,314	748,912	43,806	220,632
Ratio to Standard	0.86	0.82	0.98	0.96

Note: All ratios are with respect to results for the standard assumptions

Payments for water include capital repayment based on the City's fixed allocation. They increase from an initial $5 per acre foot to $40 by 2050, the final year for repayment. This schedule was developed to make repayment as easy as possible by postponing the larger payments to later years. Some political supporters expected that the larger future payments would never be enforced by the federal government. These scheduled payments, totaling $17 million, were deflated at 4% per year for expected long run inflation. The commodity payment for delivered water started at $53 per acre foot in 1985 dollars, and is expected to increase with inflation, therefore remaining constant in real terms.

Phase one of the treatment plant is projected to cost $79,588,350 in 1985 dollars, spread over four years. The second phase, planned for completion by 2007 is estimated to cost 60% of the initial construction cost. Operating costs have been estimated at

$41.18 per acre foot. The cost of new transmission mains total $43 million for phase one, and an additional $16 million for phase two. The total estimated cost to Tucson Water of utilizing CAP supplies, discounted at the 3.25% rate was $397 million. At the higher 8.625% rate, the discounted cost was $175 million.

10.8.5 BENEFIT-COST ANALYSIS

Before completing the benefit cost analysis, consideration was given to the possibility that less than the expected amounts of CAP water would be available in some years and that the CAP water might be inferior in quality and thus economic value to previously used groundwater. To take account of these possibilities, the benefit estimates were reduced by 20%. While this adjustment for risk is entirely appropriate, its correct size is unknown.

When CAP water was first used by Tucson Water some time after completion of this analysis, the CAP water turned out to be corrosive whereas the groundwater it replaced was high in lime. As a result, numerous older pipes began to leak, and the City was forced to replace distribution pipes and pay damages to a number of citizens. The political firestorm set off by leaking pipes and water heaters along with brown suspended solids delivered with the water to some homes forced the City to postpone use of CAP water for several years and change its treatment method, thereby incurring large unexpected costs and postponing the expected benefits.

At the preferred 8.625% discount rate, total costs over the 100 year projected life of the project are $175 million compared to benefits ranging from $36 to $115 million, depending on the choice of demand model. To obtain point estimates of benefits, the demand model results are combined, using a weight of 0.7 for the multiplicative model results and 0.3 for the linear model. These weights reflect the author's judgment regarding the probability that each model best represents reality. The weighted benefits are $91 million, yielding a benefit cost ratio of 0.52.

The original 3.25% discount rate yields discounted costs of $397 million and benefits ranging from $183 to $729 million yielding a weighted average of $565 million. The benefit cost ratio is 1.42, as shown in Table 10.5. Using the weighted benefits of the two models yields an internal rate of return of 4.86%.

The benefit model includes only the direct benefits of using CAP water. If the increased availability of water encourages immigration, landowners and developers may receive pecuniary benefits from the project at the expense of current residents who will be forced to pay higher water bills. To examine the implications of this possibility, the excess cost of the project of $84 million at the 8.625% discount rate was distributed over all new households projected to be served by Tucson Water. This resulted in a cost per new household of $2,514 in 1985 dollars. Thus current customers of Tucson Water would break even on their investment in buying and using CAP water supplies only if each new residential unit is assessed a connection fee of $2,514. This could be imposed as a surcharge over the first 5 or 10 years of service for each new customer instead of as a single lump sum payment. Without such a payment, current residents would be better off without the CAP.

Table 10.6 Benefit Cost Summary
Millions of 1985 Dollars

	Discount Rate	
	8.625%	3.25%
Cost	175	397
Exponential Model		
Benefits	115	729
Net Benefits	-60	332
Linear Model		
Benefits	36	183
Net Benefits	-139	-214
Weighted Results		
Benefits	91	565
Net Benefits	-84	168
Benefit/Cost Ratio	0.52	1.42
Internal Rate of Return, Weighted Results 4.86%		

References

Agthe, D.E., R.B. Billings and J.M. Dworkin, 1988. "Effects of Rate Structure Knowledge on Household Water Use," *Water Resources Bulletin* 24(3): 627-630.

American Society of Civil Engineers, 2001. *ASCE's 2001 Report Card for America's Infrastructure*, Washington, DC.

Billings, R.B. 1990. "A Demand Based Benefit-Cost Model of Participation in a Water Project," *Journal of Water Resource Planning and Management* 116(5): 593-609.

Billings, R.B., and W.M. Day, 1989. "Demand Management Factors in Residential Water Use: The Southern Arizona Experience," *Journal American Water Works Association* 81(3): 58-64.

Bruce, N., 1998. *Public Finance and the American Economy.* Addison-Wesley, Reading MA, USA.

Haarmeyer, D. 1992. *Privatizing Infrastructure: Options for Municipal Water-Supply Systems.* Policy Study 151, The Reason Foundation, Los Angeles.

Israel, D. K., 2001. "The Distributional Impact of Price Ceilings on Water in Developing Countries: An Analysis of Household Water Expenditures in Urban Bolivia," A paper presented at the Southern Economic Association, Tampa, Florida, November.

Levin, R.B., P.R. Epstein, T.E. Ford, W. Harrington, E. Olson, and E.G. Reichard, 2002. "U.S. Drinking Water Challenges in the Twenty-First Century," *Environmental Health Prospectives,* 110 (February, Supplement 1): 43-52.

Martin, W.M., N.H. Laney, and A.W. Griffin, 1985. *Saving Water in a Desert City*, Resources for the Future, Inc., Baltimore.

Mishan, E.J. 1973. *Economics for Social Decisions: Elements of Cost-Benefit Analysis*, Prager Publishers, Inc., New York, NY.

Neveu, R.P., 1989. *Fundamentals of Managerial Finance,* Southwestern Publishing, Cincinnati, Ohio.

Timmins, C., 2002. "Does the Median Voter Consume Too Much Water: Analyzing the Redistributive Role of Residential Water Bills" *National Tax Journal, forthcoming.*

U.S. Congressional Budget Office, 1987. *Financing Municipal Water Supply Systems.* Washington, D.C.

U.S. EPA 2001. *Drinking Water Infrastructure Needs Survey*, EPA 816-R-01-004 Washington, D.C.

U.S. National Research Council, 2000. *Watershed Management for Potable Water Supply,* Committee to Review the New York City Watershed Management Strategy, Water Science and Technology Board, Academy Press, Washington, D.C.

Water Infrastructure Network 2000. *Clean and Safe Water for the 21st Century.* American Metropolitan Water Agencies.

Whittington, D., D.T. Lauria and M Xinming, 1991. "A Study of Water Vending and Willingness to Pay for Water in Onitsha, Nigeria," *World Development,* 19(2/3): 179-198.

Wietzman, M. L. 1988. "Why the Far-Distant Future Should Be Discounted at Its Lowest Possible Rate," *Journal of Environmental Economics and Management* 36(3): 201-208.

Chapter 11

Drinking Water Disinfection in the United States: Balancing Infectious Disease, Cancer and Costs, Market and Nonmarket Failures

Ronnie Levin
Harvard University School of Public Health
Mark A.R. Kleiman
University of California, Los Angeles

11.1 Introduction

Arguably, the most important public health improvement of the 20th century is the virtual elimination of the most deadly waterborne infectious diseases, such as cholera, through the treatment of drinking water (CDCP, 1999). But current treatment practices have not reduced health risks to zero. Recent studies indicate that waterborne infectious disease (WBID) - albeit less severe - remains more widespread in the U.S. than previously thought. On the other hand, the disinfection methods used in the U.S., most commonly chlorine species, produce a variety of disinfection by-products (DBPs), some of them carcinogenic. Addressing microbial risks by increasing disinfectant doses or longer exposure times will exacerbate the DBP problem. There are solutions that address both WBID and DBPs simultaneously, such as filtration, especially microfiltration and nanofiltration, or improved distribution operations and maintenance, but they also have their advantages and disadvantages, including higher costs.

Regulators and operators of U.S. water systems face competing considerations relating to drinking water disinfection, including both risk-risk and risk-cost trade-offs,

D.E. Agthe et al. (eds.), Managing Urban Water Supply, 167–185.

complicated by substantial uncertainties, both technologically and on a risk-management basis (Levin and Kleiman, 1999). At any point, choices must be made against a fixed set of existing technical options, as well as about how much to invest in developing new technologies. Ample evidence -- sales of bottled water and home water-treatment devices, opinion surveys, etc -- suggests widespread public misgivings about tap water in the U.S. presently (Anadu and Harding, 2000). This analysis shows that from the viewpoint of consumers/voters, relatively modest investments to reduce microbial contamination of drinking water and more significant investments to reduce exposure to DBPs would be welfare enhancing, including economic considerations. Yet, water utilities are unlikely to undertake these actions voluntarily.

This chapter uses a benefit-cost approach to describe that entire complex of drinking water disinfection decisions and also discusses the institutional factors that lead publicly-owned U.S. water utilities to make choices that appear suboptimal for both themselves and their customers. All monetary valuations are 1998 dollars, unless otherwise noted.

11.2 Waterborne Infectious Disease in the United States

11.2.1 MICROBIAL CONTAMINATION OF DRINKING WATER

Hundreds of human pathogens, including bacteria, protozoa, and viruses, can contaminate drinking water. Modern drinking water disinfection practices are most effective against bacteria while contamination is most common by protozoa and viruses. Drinking water can be contaminated by microbes at many points: as source water before entering the distribution system, within the water treatment plant through backwashing filters or other "not recommended" practices, or within the distribution system, either from holes and leaks or from cross-connections usually involving the wastewater system. Contamination can also occur within a building.

Microbial contamination of drinking water is widespread in the U.S. Numerous studies have found source waters to be contaminated (Harvey et al., 1976; LeChevallier et al., 1991a and 1991b; Rose et al., 1991; Thomas et al., 1999). Samples of treated water show that pathogens can pass through conventional water treatment (Akin & Jakubowski, 1986; LeChevallier et al., 1991b; Rose et al., 1991; Hunter et al., 1998; Aramini et al., 2000) and even some filtration plants (MacKenzie et al., 1994; Goldstein et al., 1996), and can survive in the distribution system with residual chlorine present (Payment, 1999). Even pristine waters are contaminated (Ongerth, 1989; Chauret et al., 1995). Contamination of both source and treated water with human enteric viruses is also widespread (Keswick et al., 1985; Hurst, 1991).

11.2.2 ESTIMATED INCIDENCE OF WATERBORNE INFECTIOUS DISEASE

In the U.S., Federal regulations require disinfection of vulnerable public water supplies and protection of many drinking water sources. With the exception of a massive outbreak in 1993, the US also has not shown evidence of many serious WBID outbreaks over the past 50 years or so. This has resulted in the overly optimistic belief that WBID has been essentially eliminated here. Consequently, data on U.S. incidence of WBID are sparse, and no single data base presents a complete picture of it. Numerous recent epidemiological and other studies suggest, however, that WBID is an underestimated cause of mortality and substantial morbidity in the U.S. currently.

The Centers for Disease Control and Prevention (CDCP, formerly the Centers for Disease Control/CDC) maintain a passive surveillance system for waterborne disease outbreaks; participation is voluntary and data are self-reported. Data from 1970-1998 indicate that on average about 20 outbreaks of waterborne infectious disease, associated with 5,000 - 7,000 cases of illness, are documented annually; numbers of outbreaks and cases have declined in recent years (St. Louis, 1988; Levine et al., 1990; Herwaldt et al., 1991; Moore et al., 1993; Kramer et al., 1996; Levy et al., 1998; Barwick et al., 2000). An exception to this apparent trend is the massive outbreak in Milwaukee in 1993, affecting more than 400,000 people and resulting in over 4,000 hospitalizations (MacKenzie et al., 1994). It is widely assumed that the CDC data represent only a small fraction of the actual incidence of WBID, but the degree of underreporting is unknown (Frost et al., 1996).[29]

Several other categories of studies contribute to a picture of the extent of WBID in the U.S., though none does so completely. Two studies estimated the actual national incidence of WBID in the U.S. Bennett et al. (1987), using information from experts at CDC to estimate the pathogen-specific incidence of infectious disease, estimated 940,000 cases per year of WBID, associated with about 900 deaths annually. Morris and Levin (1995), also using secondary sources on pathogen-specific illnesses, estimated 7.1 million cases per year of mild-to-moderate waterborne gastrointestinal (GI) disease; 560,000 cases per year of moderate-to-severe GI disease; and about 1,200 associated deaths annually.

Some studies have assessed empirically the relationship between consumption of tap water and the occurrence of GI symptoms within a city. Two separate intervention trials by Payment and colleagues (1991 and 1997) found that consumption of tap water was significantly related to the incidence of GI symptoms, although the water met federal standards. Those studies found that 14-40% of GI illness was water-related. A study to test the effect of blinding on such intervention studies found a similar correlation with attributable risk calculated at 0.24 (Colford et al., 2002).

Even more recently, several time-series analyses evaluated the relationship between fluctuations in daily turbidity levels in drinking water (a measure of the "cloudiness" of the water) and the daily use of medical services for GI illnesses in the city (Morris et al., 1996; Schwartz et al. 1997; Schwartz et al., 2000). In all three studies, turbidity levels well within federal drinking water standards were significantly associated with increased risks for GI disease in the general population, in young children and in the elderly. While the results varied by city and population sector, these

[29]Studies have shown that only 1-10% of those suffering gastrointestinal symptoms are likely to seek medical care (Garthright et al., 1988; Perz et al., 1998). Perz also estimated that, unless there is a major epidemic, only 3 of 10,000 cases of severe cryptosporidiosis in the non-AIDS adult population that seek medical help would be reported to local health authorities, and of those, only the smallest percentage are likely to be reported to CDC.

studies suggest that 10% or more of GI medical care is related to drinking water quality. Studies on a comparable water system in Canada presented similar results (Aramini et al., 2000).

Finally, several studies have shown evidence of undetected outbreaks of cyptosporidiosis in both the general and HIV-positive populations despite state-of-the-art water treatment (for example, Morris et al., 1996 and 1998; Goldstein et al., 1996). In the Goldstein study, the water system met all federal drinking water standards.

The available data on microbial contamination of drinking water and studies of pathogen-specific diseases are consistent with even higher incidence estimates. Grubbs (et al., 1992) estimated between 400,000 and 1 million *Giardia* infections annually among the 150 million users of U.S. surface water, using the data on the microbial integrity of source water from Harvey, LeChevallier, Rose and Thomas cited above. With that same exposure data, Berger et al. (1992) estimated 10 million *Cryptosporidium* infections per year; EPA (1998a) assumes cryptosporidiosis incidence rates close to this level.

Together, these disparate studies (microbial contamination levels, empirical studies including intervention trials, and time-series analyses) contribute to a picture of widespread, albeit mild, WBID risk in the U.S. The pathogens detected (most often *Giardia* and *Cryptosporidium*) generally produce self-limiting GI illness with low case-fatality rates; the morbidity detailed in the intervention trials and time-series analyses are consistent with that. Only in extremely vulnerable populations, such as the immunocompromised, elderly or infirm, is mortality more than a tiny risk.

For our analysis, we use the Morris and Levin (1995) estimates because that study is the most complete to date and is used elsewhere (e.g., Ford, 1999; Levin et al., 2002). However, that study may underestimate the actual incidence of WBID, for several reasons.

- It includes only seven pathogens of the hundreds known to sometimes contaminate drinking water, and only includes GI illness of the infections that can be waterborne.
- It estimates disease on a pathogen-specific basis only, while CDC data show that, over the past 30 years, one third or more of the documented waterborne disease outbreaks have had an unidentified etiology (St. Louis, 1988; Levine et al., 1990; Herwaldt et al., 1991; Moore et al., 1993; Kramer et al., 1996; Levy et al., 1998; Barwick et al., 2000).
- Morris and Levin's estimates do not include the most recent findings on cryptosporidiosis, which suggest incidence estimates at least an order of magnitude higher.
- Finally, extrapolating from the attributable-risk estimates in either the Payment et al. (1991, 1997) papers or the time-series analyses would also yield far higher morbidity estimates.

On the other hand, some of the Morris and Levin (1995) data reflect conditions before the EPA regulations of 1989 began to take effect in 1993. In that case, current WBID rates may be lower than those estimated.

11.2.3 THE MONETIZED COSTS OF WATERBORNE INFECTIOUS DISEASE

The variety of diseases involved and the absence of studies that directly estimate the willingness-to-pay to avoid various levels of WBID morbidity complicate the valuation of WBID control efforts. We base our valuation on a study of the direct costs of a single giardiasis case: $3,400, which includes medical costs (direct and indirect) and loss of productivity and leisure (Harrington et al., 1991). For benefit-cost analysis, one should add to the cost-of-illness (COI) estimates a valuation of pain and suffering; the sum represents an estimate of the total willingness-to-pay (WTP) to avoid disease (EPA, 1997). This would increase the unit cost substantially, because, although giardiasis is a relatively mild disease in terms of its case-fatality rate, it tends to be long-lasting, with a mean duration of 30 days (Hill, 1985; Harrington et al., 1991).[30] In addition, only costs (lost productivity and leisure) incurred by the ill individual are included; costs to parents of young children, children of aged or debilitated parents, spouses, etc., for lost time caring for the ill person are omitted. A complete COI analysis should also include these real costs.

We treat more severe (based on case-fatality rate) but shorter-duration infections such as salmonellosis, shigellosis, and *E. coli* infections as having morbidity costs per case equivalent to giardiasis (again, strictly on a COI basis). We assign proportionately lower values for milder and typically briefer infections with *Cryptosporidium* (5-10 days [Jones 1985, MacKenzie et al., 1994]), which we value at one-quarter of the per-case cost of giardiasis, or $850 per case. Waterborne viral infections (typically lasting 1-3 days [Guerrant and Hughes, 1985]) are valued at 1/15 of the per-case cost of giardiasis, or $230 per case. By comparison, in its most recent rule making (EPA, 1998a and 1998d), EPA valued a case of cryptosporidiosis at $2,000, almost 2.5 times higher. Our WBID unit cost estimates are summarized in Table 11.1. Because waterborne infectious diseases tend to kill those weakened by age or poor health, the average number of quality-adjusted life-years lost per death is lower for WBID mortality than for industrial accidents. Estimates of willingness-to-pay to avoid death derived from wage premia in high-risk occupations (approximately $6.7 million per avoided death [Moore and Viscusi, 1988]), therefore would overstate the value of avoiding WBID deaths. In this analysis, we use a relatively low value of $2 million per death. Again for comparison, EPA (1998a) used a point estimate of $5.6 million per life, although that study's estimated case-fatality rate is a small percentage of the one reported here.

Non-health costs are also associated with WBID, especially when outbreaks are declared; these include averting behavior, such as boiling or purchasing water, and other costs to business and government to deal with the emergency. Using data from a 1983 giardiasis outbreak, Harrington et al. (1991) also estimated those costs: averting behavior -- about $3.50 per person per day for the entire population served, and about $80 per person per incident for other costs to business and government, excluding costs to the water utility itself, which were not included in Harrington's study. (Including costs to the water utility would probably more than double this estimate.) Table 11.1

[30]On average, WTP estimates are about 4 times greater than COI estimates (EPA 2002, Exhibit 5-2), with a range of 2:1 to 32:1.

presents non-health unit costs also. We ascribe these costs only to persons in the service area of a documented outbreak, using incidence estimates from the CDC data; they are dwarfed by the morbidity and mortality valuations. Our total estimate of the annual cost of WBID is about $7 billion per year as shown in Table 11.2.

Table 11.1 Unit Cost Estimates for Waterborne Infectious Disease (1998 dollars)

	Unit Cost	Notes
Case of waterborne gastrointestinal illness (Harrington, 1991)	$3,400[31]	Based on giardiasis case; includes direct & indirect medical costs and productivity & leisure losses
Averting behavior (Harrington, 1991)	$3.50/day/ person	Midpoint of 3 costed scenarios
Costs to business & government (Harrington, 1991)	$80/ person	Excludes costs to the water utility itself
Mortality (Moore & Viscusi, 1988)	$2 million	Lower range of estimates

11.3 Disinfection Byproducts

Chlorine reacts with organic contaminants found in most surface waters to form disinfection byproducts (DBPs). The trihalomethanes (THMs), including choloroform, were the first DBPs discovered, and are the most commonly occurring. Several hundred DBPs have already been identified, and many remain to be identified (Singer, 1993).

Residual levels of chlorine and some other disinfectants are maintained in drinking water until it reaches the consumer's tap, to prevent recontamination or microbial regrowth within the water distribution system. These levels are believed to be safe, but higher levels may create health risks in addition to giving the water an unpleasant taste and odor. There are no published estimates of such risks. The current craze for bottled water (now over $3 billion/year, International Bottled Water Association, 1996) may reflect consumers' dissatisfaction with the flavor and odor of their tap water or concern about its safety (Anadu and Harding, 2000), as well as reflecting lifestyle statements.

[31]Midpoint of 2 'more reasonable' scenarios; Scenario C is omitted from the calculation because it underestimates losses by ascribing no value to time losses for those not currently active in the workforce, such as retirees and homemakers; no value is ascribed to children's illnesses.

Table 11.2 Estimated Annual Monetized Costs of Waterborne Disease (1998 dollars)

Outcome	Unit Cost	Incidence	Total	Notes
Morbidity				
Moderate-severe Illness	$3,400[a]	560,000[b]	$1.9 billion	includes medical & lost productivity costs to individuals from a giardiasis outbreak; may underestimate the cost of
Giardiasis	$3,400[a]	260,000[b]	$880 million	includes medical & lost productivity costs to individuals from a giardiasis outbreak
Cryptosporidiosis	$850	420,000[b]	$360 million	unit cost is 1/4 of giardiasis unit cost, based on duration of illness
Viral illnesses	$230	6.5 million[b]	$1.5 billion	unit cost is 1/15 of giardiasis unit cost, based on duration of illness
Mortality	$2 mil	1,200[b]	$2.4 billion	
Averting behavior	$3.50/day/ person[a]	23 outbreaks[c]	$11 million	only for documented outbreaks; assumes average system=5,000 people & average outbreak= 30 days
Costs to business & government	$80 person[a]	23 outbreaks[c]	$ 9 million	only for documented out-breaks; assumes average system=5,000 people & average outbreak= 30 days; excludes costs to water utility
Total			**$7.0** billion	excludes costs to water utility itself

Source: a. Harrington et al., 1991; b. Morris and Levin, 1995; c. Centers for Disease Control surveillance data

11.3.1 TOXICOLOGY AND EPIDEMIOLOGY OF DBPs

Overall, limited research has been done on the health significance of most DBPs. The extant research indicates that some DBPs, including THMs, are biologically active in *in vitro* tests and four DBPs are classified as probable human carcinogens (B2 carcinogens) using evidence of carcinogenicity in animals: chloroform, bromoform and bromodichloromethane (all trihalomethanes), and dichloroacetic acid (EPA 1994b). A few studies suggest there may also be other health effects.

Epidemiological studies of the association between consumption of chlorinated water and cancer have been inconsistent. Morris (et al. 1992) conducted a meta-analysis that found a statistically significant relationship between bladder and rectal cancers, and exposure to disinfected drinking water. That analysis was quite controversial (e.g., Cantor, 1994). Other studies have found an association between cancers of the bladder, colon and rectum, and various measures of chlorination by-product contamination, although the results are not consistent (e.g., summary in Murphy and Craun, 1990; Vena et al., 1993; Koivusalo et al., 1994; Doyle et al., 1997; Freedman et al., 1997, Cantor et al., 1998; Hildesheim et al., 1998). Other sites have not been ruled out (Koivusalo et al., 1995; Cantor et al., 1999), especially gastrointestinal targets (e.g., Ijsselmuiden et al., 1992; Dunnick and Melnick, 1993; Cantor, 1997). Data from animal toxicological studies investigating the mechanisms of DBP mutagenicity and carcinogenicity have confirmed metabolic changes consistent with carcinogenicity (George et al., 2000) and have demonstrated factors that may explain the organ-specificity of DBP carcinogenicity (Landi et al., 1999). EPA (1998c) concluded that a causal relationship exists between exposure to DBPs and bladder cancer.

There is also some evidence of other health effects (reproductive, developmental, immune-system and other non-cancer effects) of DBPs at current exposures, but they have not been conclusively demonstrated (Bove et al., 1995; Swan et al., 1998; Waller et al., 1998; Klotz and Pyrch, 1999; French et al., 1999). Nonetheless, the risks are probably not zero.

11.3.2 ESTIMATED INCIDENCE OF HEALTH EFFECTS ASSOCIATED WITH EXPOSURE TO DBPs

The relative risk for excess bladder and rectal cancers associated with exposure to chlorinated drinking water calculated in the Morris (et al., 1992) meta-analysis is 1.15 (the 95% confidence interval is between 1.09 and 1.20); pooled organ-specific attributable risk estimates were 4,200 bladder cancer cases and 6,500 rectal cancers per year. This is not inconsistent with an earlier analysis that estimated an attributable risk of 28% for non-smokers (Murphy and Craun, 1990). Similarly, EPA (1998b and 1998c) estimated 1,100-9,300 bladder cancers/year attributable to DBP exposure. For our study, we use the Morris (et al. 1992) attributable risk estimates because the methodology has been presented most clearly and because the estimates are consistent with other analyses.

11.3.3 THE MONETIZED COSTS OF EXPOSURE TO DBPs

Monetizing the reduction in carcinogenic risk is complicated by the latency between exposure and manifestation. First, the lag means that those who develop cancer are generally much older than those who are exposed, reducing both the number of life-years lost and their average quality. It also means that the damage should be discounted because it arrives in the future.

One study, designed specifically to estimate the willingness-to-pay for avoiding the risks associated with THM contamination of drinking water, found a WTP of $270,000 to avoid a death 30 years in the future (Mitchell and Carson, 1986). The difference between this and the much higher Moore and Viscusi (1988) estimate ($6.7 million) cited above likely represents some combination of reduced valuation on later death due to the reduced loss of quality-adjusted life-years and discounting to present value.[32] Multiplying the Mitchell and Carson (1986) value ($270,000) by the attributable risk estimate from Morris et al. (1992) (10,700 estimated bladder and rectal cancers annually) and assuming mortality approaching unity (unlikely) would give an estimate due to lives lost of about $2.9 billion per year. Of course, assuming some fraction of these cancers is not fatal reduces this estimate. For instance, assuming that three-quarters are non-fatal and valuing non-fatal cases at 10% of fatal cases (i.e.: $27,000) would reduce the damages to about $1 billion per year.

For comparison, EPA's regulatory analysis (EPA, 1998b and 1998e) used a unit cost estimate of $5.6 million per fatal bladder cancer and estimated a willingness-to-pay to avoid a non-fatal bladder cancer of about $590,000. EPA also estimated that 23% of bladder cancers are fatal and 77% are not. Combining this unit cost algorithm with an estimate that DBPs in drinking water could be responsible for up to 9,300 bladder cancers per year resulted in an EPA estimate of about $16 billion/year in damages.[33]

Because of the uncertainties in both estimates, we present the range of $1-16 billion per year for these damages (Table 11.3). Potential costs associated with other cancers and with non-cancer health damage are omitted here. While no estimates are available, the expected values are certainly not zero.

11.4 Water Treatment Operations and Constraints

Surface water is treated in various ways, including disinfection, before it gets to the consumer. Surveillance data collected by the Centers for Disease Prevention and Control (St. Louis, 1988; Levine et al., 1990; Herwaldt et al., 1991; Moore et al., 1993; Kramer et al., 1996; Levy et al., 1998; Barwick et al., 2000) identify the attributable deficiency for all documented WBID outbreaks. Since 1971, 40-50% of the outbreaks have been associated

[32]Assuming that willingness-to-pay to avoid a current cancer death occurring at a typical age for such a death is $2 million (the same valuation used here for a death resulting from WBID), the implied (real) discount rate is approximately 7% -- much higher than risk-free returns but not unreasonable as a real discount rate for consumers.

[33]EPA estimated about 24% ($4 billion/year) of DBP-related health effects could be avoided by water treatment. It is this estimate that is presented in both the published rule and the Regulatory Impact Analysis.

with inadequate water treatment, and an additional 15-20% has been associated with deficiencies within the water distribution system.

Table11.3 Estimated Annual Monetized Costs of Cancers Associated with Exposure to Disinfected Water Alternative Unit Costs (1998 dollars)

Outcome	Unit Cost	Attributable Incidence	Total Cost	Notes
bladder & rectal cancers	$270,000[a]	10,700[b]	$2.9 billion	assumes all are fatal; WTP estimate is discounted
bladder & rectal cancers	$270,000[a] $27,000	10,700[b]	$940 million	assumes 3/4 are non-fatal and that morbidity costs are 10% of fatal costs (both discounted)
bladder cancers	$5.6 mil[c] $590,000[c]	9,300[c]	$16.2 billion	assumes 3/4 are non-fatal; fatal cases costed at $5.6 million; non-fatal cases costed at $590,000
Total DBP Risk of Cancer			**$1-16 billion**	

Source: a. Mitchell and Carson, 1986; b. Morris et al., 1992; c. EPA 1998e.

Although numerous other disinfection techniques and practices are available, chlorine species are by far the most commonly used chemical disinfectant method in the U.S. Many factors determine the rate of disinfection, including disinfectant dosage, time of contact, chlorine demand, water temperature and quality, and the nature of the organisms to be destroyed. Of these, dosage and contact time are particularly relevant to this discussion. Contact time is the interval between the addition of a chemical and delivery of the treated water to the user inactivation of the chemical through reaction. Increasing either dosage or contact time reduces infectious disease risk. However, these same two factors, along with the concentration of organic contaminants when the disinfectant is added, also determine the levels of disinfection by-products formed. Increasing the dose of chlorine (or other chemical disinfectant) may also have other direct adverse health impacts because all chlorine species are at least moderately toxic to humans (Sax and Lewis, 1989), and also risks giving the water an unpleasant taste and smell.

Another difficulty in the management of microbial contamination of drinking water is its sporadic nature. Levels of pathogens can change by several orders of magnitude within minutes. Water systems must maintain levels of disinfection adequate for the most severe conditions they anticipate.

Given restrictions on desirable levels of both disinfectants and DBPs, current best practice favors a "multiple barriers" approach, with numerous independent treatment steps. In the search for lowered health risks, many options are available, including pre- and post-disinfection; multiple types of filtration (to remove both pathogens and DBP precursors); changes in coagulation, flocculation, and sedimentation processes, again to remove precursors to DBP formation; temporary closing of inlets when water quality changes occur; distribution system changes; and use of alternative or complementary disinfecting agents such as ozone and chloramine. Source water protection and rigorous distribution system maintenance and operations are also important. Each option has its own benefits, risks, operational constraints, and costs. But decreased exposure to DBPs and WBID risk need not occur at each other's expense.

11.5 EPA's Recent Microbial Contamination Rule

In 1989, EPA issued regulations requiring disinfection of all water systems supplied by surface water, plus filtration treatment in most surface water and some groundwater systems (54 FR 27486; June 29, 1989). EPA simultaneously established microbial standards and monitoring requirements that apply to all community and non-community water systems (54 FR 27544; June 29, 1989). Both regulations became effective June 1993. The goal of the disinfection rule was to "ensure less than one case of microbiologically caused illness per year per 10,000 people" (10^{-4}), focusing on waterborne *Giardia* infection.

EPA acknowledged "shortcomings" in the regulations shortly after publication and subsequently proposed to strengthen them (EPA, 1994a). At the same time, EPA proposed to tighten the disinfection by-product rule (EPA, 1994b) and proposed a related monitoring and reporting requirement rule (The Information Collection Rule; EPA, 1994c). All three rules were developed through an extensive regulatory negotiation that included representatives of federal and state government, the water utilities, consumer advocates and other key actors. The Information Collection Rule has produced important data for the final rules.

The final Interim Enhanced Surface Water Treatment Rule (EPA, 1998a) addresses microbial contamination. It is designed to reduce the likelihood specifically of cryptosporidiosis, although the goal of achieving a 10^{-4} incidence rate (for giardiasis or any other WBID) has been abandoned. Using several alternative scenarios, the EPA analysis estimates that the new rule will prevent 110,000 to 463,000 cases of cryptosporidiosis annually. The estimated cost of the rule is about $307 million/year for water systems serving more than 10,000 people each and using surface water or vulnerable groundwater. This is more than a 30% decrease in estimated costs from the proposal (EPA, 1994a); the decrease resulted from data that became available after the 1994 proposals. These cost estimates were developed in partnerships with the water industry established during the regulatory negotiation. The relatively low cost of meeting the new standards in large water systems dominates the *average* cost estimate. The marginal unit cost of extending the coverage to small, largely rural systems will be substantially higher than the average.

Table 11.2 summarizes our estimated annual monetized health costs for WBID in the U.S., which are about $7 billion/year. Table 11.4 compares estimated annual benefits and costs; benefits far exceed costs. According to our calculations, reducing overall WBID incidence as little as 5% would make the regulation cost-justified; the actual results could hardly be so modest. EPA (1998d) estimated that 53-72% of cryptosporidiosis cases were likely to be avoided by this rule; if actual WBID damages were reduced by 60%, the benefits of the rule would be 14 times greater than its costs.

Table 11.4 Estimated Annual Monetized Benefit:Cost Ratios for Reducing Waterborne Disease and Cancers Associated with Exposure to Chlorinated Water (1998 dollars)

	Waterborne Disease	Cancers/Disinfected Water
Medical & direct costs (benefits or damages)	$7.0 billion	$1-16 billion
Water treatment costs[a]	$307 million[b]	$701 million[c]
Percent reduction in health damage to break even[d]	<5%	4-70%
Estimated reduction in damages likely with water treatment	60%[e]	24%[f]
Benefits:costs at likely implementation level	14:1	0.33-6:1

a. All treatment cost estimates were developed through the regulatory negotiation convened by EPA, which included representatives of federal and state government, the water industry, consumer advocates and other key actors.

b. 63 FR 69477; December 16, 1998. *Interim Enhanced Surface Water Rule*; only includes costs to water systems serving >10,000 and using surface water.

c. 63 FR 69389; *Disinfectants and Disinfection Byproducts Rule*; only includes water systems required to disinfect.

d. How much damages would need to be reduced for benefits to equal costs

e. EPA 1998d estimated 53-72% reduction in cases of cryptosporidiosis; consistent with CDC surveillance data from 1971-1998.

f. EPA 1998e estimated 24% reduction in DBP levels; this assumes linearity between DBP levels and health effects.

The benefit-to-cost ratio would be larger and the break even level of reduction in morbidity would be lower if the following were used in these estimates.

- Extrapolate from the Payment et al.(1991, 1997) trials or the time-series studies of Schwartz et al. (1997), Schwartz et al. (2000) and Morris et al. (1996)
- Use more recent estimates for cryptosporidiosis incidence, including estimates for outbreaks with an unidentified etiology
- Use EPA's unit benefit estimates for WBID incidence or related mortality
- Use willingness-to-pay rather than cost-of-illness values for morbidity

Conversely, a lower current baseline of WBID due to previous improvements not reflected in the published literature would decrease the break-even point.

11.6 EPA's Recent Disinfection Byproducts Rule

When EPA proposed to strengthen the drinking water disinfection rule, it simultaneously proposed to tighten the DBP regulation (EPA, 1994b). This proposal, also developed through the regulatory negotiation process with key affected entities, had two phases: Stage 1 (based on available data and tightening existing standards only slightly, thereby creating relatively little dislocation in the industry) and Stage 2 (potentially imposing a considerably tighter standard). When the rule was actually published (EPA, 1998b), only the benefits and costs of Stage 1 were estimated monetarily; the final estimated cost, $701 million per year, is only two-thirds of the cost estimated in the proposal (EPA, 1994b), again due to the availability of new data.

Table 11.3 summarizes our estimated annual monetized health damages for exposure to DBPs. We present estimates for DBP-related damages that range from $1-16 billion, depending upon the assumptions used. Table 11.4 compares estimated costs and benefits. EPA (1998b) estimates that DBP levels will be reduced by 24% by the new rule; if health effects are proportional to DBP levels, the rule would likely be cost-justified on these health damages alone. However, these estimates are quite controversial and some believe they may overestimate risks. The uncertainties surrounding this part of our analysis are too great overall to form an iron clad economic argument for the rule using existing risk estimates.

As Table 11.4 shows, estimated benefits of reducing exposure to DBPs through the new rule are between one-third and six times the costs, reflecting the limitations of existing data and the conservatism of conventional benefit-cost methods. A rational consumer may prefer water with less residual chlorine and DBPs, at a somewhat higher price, than most water utilities currently deliver given any of the following.

- Attributing a non-zero value to the risk of cancers at other sites
- Putting a higher valuation on a cancer death or on a non-fatal cancer (e.g., EPA 1998b)
- Applying a discount rate lower than the implicit 7% real rate we used in the valuation of a future cancer death
- Willingness to pay for water with less unpleasant taste

On the other hand, if the epidemiological studies overestimate actual risk, total benefits will be lower. As with the microbial rule, relatively low unit costs for large water systems dominate the average-cost estimates. The per-person costs imposed by extending the rule to small systems are higher.

11.7 Market and Government Failures in Regulating Drinking Water

Unless there are significant errors in our estimates of the direct costs of waterborne infectious disease, the current costs of WBID in the U.S. (estimated at about $7 billion/year) dwarf the costs of water treatment. There is a very strong economic argument for increasing expenditures to reduce infectious disease risk without increasing cancer risk, and an uncertainty-laden but still highly likely economic case for controlling DBP exposure here, as well. Indeed, other analyses using very different methods have come to the same conclusions (Havelaar et al., 2000).

Why, then, does EPA need to push the water utilities to do so? Why does an industry consisting largely of public agencies need to be federally regulated at all?

First, the customers of water systems tend to be unaware either of the overall level of waterborne infectious disease or of how their own supplier compares to others. (The reader is encouraged to check this assertion on a sample of one by introspection.) Costs and aesthetic properties - taste, odor, color, and clarity - are more evident than microbial content; unless they are traveling abroad, few Americans will blame their intestinal ills on the drinking water. So, the willingness-to-pay for safe water might remain largely latent, even if water were sold in competitive markets.

Of course, in reality, consumers have little opportunity to express their choices through the marketplace, because drinking water at the tap is sold by geographic monopolies. Their principal venue for expressing dissatisfaction is buying bottled water or home filtration devices, which consumers are doing at an exploding rate. By volume, consumption of bottled water has increased more than tenfold over the past 20 years, and annual bottled-water expenditures are now well over $3 billion (International Bottled Water Association, 1996). Spending for "residential point-of-use devices" (home water filters and the like) was estimated at nearly $2 billion per year in 1995 (Water Quality Association, 1996). Although the motivations for these actions probably extend beyond concerns about WBID and DBPs, these are explicit statements of the willingness of a part of the public to pay for perceived improved water quality (Anadu and Harding, 2000).

Even if the marketplace gives water customers limited power, politics gives them another chance to express their desires, because most Americans get their water from agencies of local government. Realistically, however, it is difficult to imagine the microbial content of the local water forming a significant issue in a mayoral election. The technical complexity of the problem is formidable by the metrics of political campaigning, and the difficulty in attributing particular bad outcomes (major outbreaks aside) to water quality reduces the political salience of the potential issue nearly to zero. So consumers turn out to need protection by federal regulators even from water companies they in effect own.

11.8 Technological Innovation

Given the great potential for improving how drinking water is handled, why the glacial pace of technological innovation in water treatment and delivery? In all areas, regulatory practice tends to accept the constraints of existing technology as "given;" to require the use of the "Best Available Technology" seems to be as far as a regulator can push. By both practice and public acceptance, regulatory benefit-cost analysis, too, is limited to existing (or about-to-be-available) technologies for which cost and effectiveness estimates exist. There is no consensus about whether, when, and how regulators should attempt to "force" technological innovation, and attempts to do so attract vigorous resistance from regulated industries.

From the consumers' perspective, this analysis suggests that the U.S. drinking water industry may be under-investing in technological innovation for water treatment. Modest additional investments in water treatment might substantially improve the terms of both the risk-cost and risk-risk tradeoffs.

Who would make such investments? The overall level of innovation is a public good among the drinking-water companies, so there would still be a collective-action problem even if the benefits to consumers of safer water were completely internalized (economically or politically) in the decision processes of the water companies. Vendors of equipment or reagents might make such investments, but process optimization is difficult to patent, and some of the "new" technologies worth exploiting may in fact be already known, such as process improvements or combinations of disinfectants.

The incentives of individual systems and their managers argue strongly against being pioneers or early adopters of new technologies. Consumer tolerance for a few more years of drinking water carrying an annual risk of *Giardia* infection of 10^{-3} rather than 10^{-4} is certainly much higher than consumer tolerance for even a day's interruption in drinking-water supply. System managers optimizing for their own consumers will underinvest in process innovation because its benefits are nonexcludable and, therefore, it makes sense to "free ride". Late adopters get most of the benefits of the innovation with very little of the risks and costs of "shaking down" new processes. Principal-agent problems as between managers and consumers - with the possible losses from innovation much more visible than the potential gains - will tend to exacerbate the problem. These factors suggest that the market will not efficiently provide technological innovation.

11.9 Conclusions

We estimated that waterborne infectious diseases impose annual costs on U.S. households of about $7 billion. For an annual expenditure of less than 5% of that amount, these losses could be reduced substantially. There also appears to be a strong but more qualified case for increased funding to reduce exposure to the carcinogenic by-products of drinking-water disinfection in the U.S. Therefore, from the viewpoint of the consumers who bear the risk and pay the water bills, increased expenditures to improve drinking water quality could provide large net benefits. Tables 11.2 and 11.3 summarize our estimated monetized costs of waterborne disease and cancer risk associated with exposure to chlorinated water for the U.S. annually. Table 11.4 shows the range of estimated benefits and costs, using the water treatment cost estimates developed through a regulatory negotiation convened by EPA.

It is unlikely that either market forces or political forces at the local level will generate pressures for optimal expenditures to improve drinking-water quality, or for appropriate investments in the development of new technologies. This suggests that Federal regulation of local water systems, including publicly owned systems, can be welfare-enhancing. New institutional arrangements may be required to increase the rate of technological innovation in drinking-water treatment.

Acknowledgements

The authors acknowledge the insights and technical comments of Winston Harrington (Resources for the Future), Michael Robin Collins (University of New Hampshire) and Scott Summers (University of Cincinnati).

References

Akin EW, Jakubowski W., 1986. Drinking water transmission of giardiasis in the United States. *Wat Sci Tech*, 18(10): 219-226.

Anadu EC, Harding AK., 2000. Risk perception and bottled water use. *J Am Water Works Assoc*, 92 (11): 14, 82-92.

Aramini J, McLean M, Wilson J, Holt J, Copes R, Allen B, Sears W., 2000. Drinking water quality and health care utilization for gastrointestinal illness in Greater Vancouver. *Health Canada*; October 30.

Barwick RS, Levy DA, Craun GF, Beach MJ, Calderon RL., 2000. Surveillance for waterborne-disease outbreaks B United States, 1997-1998. In: *CDC Surveillance Summaries*, May 26, *Morb Mort Wkly Repts*; 49 (No. SS04): 1-35.

Bennett JV, Holmberg SD, Rogers MF, Solomon SL., 1987. Infectious and parasitic diseases. In: *Closing the Gap: The Burden of Unnecessary Illness* (ed: Amler and Dull); Oxford University Press; New York, Oxford:102-114.

Berger PS, Regli S, Almodovar L., 1992. Cryptosporidium control in drinking water. In: *Proceedings of the 1992 American Water Works Association Annual Conference*, Canada: 845-864.

Bove FJ, Fulcomer MC, Bove Klotz JB, Esmart J, Dufficy EM, Savrin JE.,1995. Public Drinking Water Contamination and Birth Outcomes. *Am J Epidemiol* 141 (9): 850-862.

Cantor KP. Water chlorination, mutagenicity and cancer epidemiology., 1994. *Am J Pub Health*, 84(8): 1211-1213.

Cantor KP.,1997. Drinking water and cancer. *Cancer Causes Cont*, 8: 292-308.

Cantor KP, Lynch CF, Hildesheim ME, Dosemeci M, Lubin J, Alavanja M, Craun G., 1998. Drinking water source and chlorination byproducts. I. Risk of bladder cancer. *Epidemiol*, 9(1): 21-28.

Cantor KP, Lynch CF, Hildesheim ME, Dosemeci M, Lubin J, Alavanja M, Craun G., 1999. Drinking water source and chlorination byproducts in Iowa. III. Risk of brain cancer. *Am J Epidemiol*, 150(6): 552-560.

CDCP (Center for Disease Control and Prevention), 1999. Ten great public health achievements United States, 1900-1999. *Morbid Mortal Wkly Rpt*, 48(12): 241-243.

Chauret C, Armstrong N, Fisher J, Sharma R, Springthorpe S, Sattar S.,1995. Correlating Cryptosporidium and Giardia with microbial indicators. *J Am Water Works Assoc*, 87(11): 76-84.

Colford JM, Rees JR, Wade TJ, Khalakdina A, Hilton JF, Ergas IJ, Burns S, Benker A, Ma C, Bowen C, Mills DC, Vugia DJ, Juranek DD, Levy DA, 2002. Participant blinding and gastrointestinal illness in a randomized, controlled trial of an in-home drinking water system. *Emerging Infectious Diseases* 8 (1), January. Available at www.cdc.gov/ncidod/EID/vol18no1?00-0481.html.

Doyle TJ, Zheng W, Cerhan JR, Hong C-P, Sellers TA, Kushi LH, Folsom AR, 1997. The association of drinking water source and chlorination by-products with cancer incidence among post-menopausal women in Iowa: a prospective cohort study. *Am J Public Health* 1997, 87(7): 1168-1176.

Dunnick JK, Melnick RL, 1993. Assessment of the carcinogenic potential of chlorinated water: experimental studies of chlorine, chloramine and trihalomethanes. *J Natl Cancer Inst*, 85(10): 817-822.

EPA, 1994a. Proposed Enhanced Surface Water Treatment Rule. 59 *FR* 38832 (July 29).

EPA, 1994b. Proposed Disinfectants and Disinfection By-product Rule. 59 FR 38668 (July 29).

EPA, 1994c. Proposed Monitoring Requirements for Public Drinking Water Supplies. 59 *FR* 6332 (Feb. 10).

EPA, 1997. *Guidelines for Preparing Economic Analyses*. Office of Policy, Planning and Evaluation; Washington, D.C.

EPA, 1998a. Interim Enhanced Surface Water Treatment Rule. 63 *FR* 69477 (December 16).

EPA, 1998b. Disinfectants and Disinfection Byproducts Rule. 63 *FR* 69389 (December 16).

EPA, 1998c. *Quantification of Bladder Cancer Risk from Exposure to Chlorinated Surface Water*. Office of Water; Office of Science and Technology; Washington DC (November 9).

EPA, 1998d. *Regulatory Impact Analysis for the Interim Enhanced Surface Water Treatment Rule*. Prepared by the Cadmus Group and SAIC for EPA Office of Water; US EPA contract 68-C6-0059.

EPA, 1998e. *Regulatory Impact Analysis for the Stage 1 Disinfectants/Disinfection Byproducts Rule*. Prepared by the Cadmus Group and SAIC for EPA Office of Water; US EPA contract 68-C6-0059.

EPA, 2002. *Draft Handbook for Non-Cancer Health Effects Valuation*. Available at: www.epa.gov/osp/spc.

Ford TE, 1999. Microbiological safety of drinking water: United States and global perspectives. *Envir Health Persp*, 107 (Supplement 1): 191-206.

Freedman DM, Cantor KP, Lee NL, Chen LS, Hei HH, Ruhl CE, Wang SS, 1997. Bladder cancer and drinking water: a population-based case-control study in Washington County, Maryland (United States). *Cancer Causes Control*, 8(5): 738-744.

French AS, Copeland CB, Andrews D, Williams WC, Riddle MM, Luebke RW, 1999. Evaluation of the potential immunotoxicity of bromodicholoromethane in rats and mice. *J Toxicol Environ Health A*, (Mar 12): 56(5): 297-310.

Frost FJ, Craun GF, Calderon RL, 1996. Waterborne disease surveillance. *J Am Water Works Assoc*, 88(9): 66-75.

Garthright WE, Archer DL, Kvenberg JE, 1988. Estimates of incidence and costs of intestinal infectious diseases in the United States. *Public Health Rpts*, 103(2): 107-115.

George SE, Nelson GM, Swank AE, Brooks LR, Bailey K, George M, DeAngelo A, 2000. The disinfection by-products dichloro-, dibromo-, and bromochloroacetic acid impact intestinal microflora and metabolism in Fischer 344 rats upon exposure in drinking water. *Toxicol Sci*, 56(2- August): 282-289.

Goldstein ST, Juranek DD, Ravenholt O, Hightower AW, Martin DG, Mesnik JL, Griffiths SD, Bryant AJ, Reich RR, Herwaldt BL, 1996. Cryptosporidiosis: An outbreak associated with drinking water despite state-of-the-art water treatment. *Annals Int Med* (March), 124(5): 459-468.

Grubbs WD, Macler BA, Regli S, 1992. *Simulation of microbial occurrence, exposure and health risks after drinking water treatment processes*. Prepared for EPA Office of Drinking Water; EPA Report 811-B-92-005; Washington, DC.

Guerrant RL and Hughes JM, 1985. Nausea, vomiting, and noninflammatory diarrhea. In: *Principles and Practice of Infectious Diseases*; Mandell, Douglas, Bennett, editors; John Wiley and Sons; New York: 646-655.

Harrington W, Krupnick AJ, Spofford WO, 1991. *Economics and Episodic Disease*. Resources for the Future; Washington, DC.

Harvey S, Greenwood JR,Pickett MJ, Mah RA, 1976. Recovery of Yersinia enterocolitica from streams and lakes of California. *Appl Envir Microbiol*, 32(3): 352-354.

Havelaar AH, De Hollander AEM, Teunis PFM, Evers EG, Van Kranen HJ, Versteegh JFM, Van Koten JEM, Slob W, 2000. Balancing the risks and benefits of drinking water disinfection: disability adjusted life-years on the scale. *Environ Health Persp*, 108(4): 315-321.

Herwaldt BL, Craun GF, Stokes SL, Juranek DD, 1991. Waterborne disease outbreaks, 1989-1990. In: CDC Surveillance Summaries, December *Morb Mort Wkly Repts*, 40 (No. SS-3): 1-22.

Hildesheim ME, Cantor KP, Lynch CF, Dosemeci M, Lubin J, Alavanja M, Craun G, 1998. Drinking water source and chlorination byproducts. II. Risk of colon and rectal cancers. *Epidemiol*, 9(1): 29-35.

Hill DR, 1985. Giardia lamblia. In: *Principles and Practice of Infectious Diseases*; Mandell, Douglas, Bennett, editors; John Wiley and Sons; New York: 1552-1556.

Hunter PR, Quigley C, 1998. Investigation of an outbreak of cryptosporidiosis associated with treated surface water finds limits to the value of case control studies. *Commun Dis Public Health*, 1 (4- Dec.): 234-238.

Hurst CJ, 1991. Presence of enteric viruses in freshwater and their removal by the conventional drinking water treatment process. *Bull World Health Org*, 69 (1): 113-119.

Ijesselmuiden CB, Gaydos C, Feighner B, Novakoski WL, Serwadda D, Caris LH, Viahov D, Comstock G, 1992. Cancer of the pancreas and drinking water: a population-based case-control study in Washington County, Maryland. *Am J Epidemiol*, 136(7): 836-842.

Immerman FW, 1997. *1986 Survey of Community Water Systems*. Report prepared by Research Triangle Institute for US EPA, Office of Drinking Water; Research Triangle Park, NC; report RTI/7805/02-02F.

International Bottled Water Association, 1996. Sales and market data from: Beverage Marketing Corporation, *Bottled Water in the US;* Alexandria, Virginia.

Jones TC, 1985. Other protozoa, including Cryptosporidium, Sarcocystis, Isospora, and Balantidium coli. In: *Principles and Practice of Infectious Diseases*; Mandell, Douglas, Bennett, editors; John Wiley and Sons; New York: 1560-1562.

Keswick BH, Satterwhite TK, Johnson PC, DuPont HL, Secor SL, Bitsura JA, Gary GW, Hoff JC, 1985. Inactivation of Norwalk Virus in drinking water by chlorine. *Appl Environ Microbiol* 51 (Aug;): 261-264.

Koivusalo M, Jaakkola JJK, Vartiainen T, Hakulinen T, Karjaleinen S, Pukkala E, Tuomisto J, 1994. Drinking water mutagenicity and gastrointestinal and urinary tract cancers: an ecological study in Finland. *Am J Public Health*: 84(8): 1223-1228.

Koivusalo M, Vartiainen T, Hakulinen T, Pukkala E, Jaakkola JJK, 1995. Drinking water mutagenicity and leukemia, lymphomas, and cancers of the liver, pancreas and soft tissue. *Arch Environ Health*, 50(4): 269-275.

Klotz JB, Pyrch LA, 1999. Neural tube defects and drinking water disinfection by-products. *Epidemiol*, 10(4-July): 383-390.

Kramer MH, Herwaldt BL, Craun GF, Calderon RL, Juranek DD, 1996. Waterborne disease: 1993 and 1994. *J Am Water Works Assoc*, 88(3): 66-80.

Landi S, Hanley NM, Warren SH, Pegram RA, DeMarini DM, 1999. Induction of genetic damage in human lymphocytes and mutations in Salmonella by trihalomethanes: role of red blood cells and GSTT1-1 polymorphism. *Mutagenesis*, 14(5- Sept.): 479-482.

LeChevallier MW, Norton WD, Lee RG. 1991a. Occurrence of Giardia and Cryptosporidium spp. in surface water supplies. *Appl Environ Microbiol*, 57 (9): 2610-2616.

LeChevallier MW, Norton WD, Lee RG. 1991b. Giardia and Cryptosporidium spp. in filtered water supplies. *Appl Environ Microbiol*, 57 (9): 2617-2621.

Levin R, Kleiman MAR. 1999. Drinking water treatment: balancing infectious disease, cancer and cost. In: *The Substance of Public Policy*, ed: SS Nagle, Nova Science Publishers, Inc; New York: 117-137.

Levin R, Epstein P, Ford TE, Harrington H, Olson E, Riechard E, 2002. US drinking water challenges in the 21st century. *Envir Health Persp Annual Review* 110 (Suppl 1): 43-52.

Levine WC, Stephenson WT, Craun GF, 1990. Waterborne disease outbreaks, 1986-1988. In: CDC Surveillance Summaries, March 1990. *Morb Mort Wkly Repts*, 39 (No. SS-1): 1-14.

Levy DA, Bens MS, Craun GF, Calderon RL, Herwaldt BL, 1998. Surveillance for waterborne-disease outbreaks United States, 1995-1996. In: CDC Surveillance Summaries, December 11, *Morb Mort Wkly Repts*; 47(5): 1-34.

MacKenzie WR, Hoxie NJ, Proctor ME, Gradus MS, Blair KA, Peterson DE, Kazmierczak JJ, Addiss DG, Fox KR, Rose JB, Davis JP, 1994. A massive outbreak in Milwaukee of Cryptosporidium infection transmitted through the public water supply. *New Eng J Med*, 331: 161-167.

Mitchell RC, Carson R, 1986. *Valuing Drinking Water Risk Reductions using Contingent Valuation Methods: A methodological study of risks from THM and Giardia*. Resources for the Future; Washington, DC.

Moore AC, Herwaldt BL, Craun GF, Calderon RL, Highsmith AK, Juranek DD, 1993. Surveillance for waterborne disease outbreaks -- United States, 1991-1992. In: CDC Surveillance Summaries, November 1993. *Morb Mort Wkly Repts*, 42 (No. SS-5): 1-22.

Moore MJ, Viscusi WK, 1988. Doubling the estimated value of life: results using new occupational fatality data. *J Policy Anal Mgmt*, (3): 476-490.

Morris RD, Audet A-M, Angelillo IF, Chalmers TC, Mosteller F, 1993. Chlorination, chlorination by-products, and cancer: a meta-analysis. Am J Public Health1992; 82(7): 955-963. Also, Erratum. 1993. *Am J Public Health*, 83(9): 1257.

Morris RD., and R. Levin, 1995. Estimating the incidence of waterborne infectious disease related to drinking water in the United States. In: *Assessing and Managing Health Risks from Drinking Water Contamination: Approaches and Applications* (Reichard and Zapponi, editors); IAHS Press (International Association of Hydrological Sciences); Great Britain: 75-88.

Morris RD, Naumova EN, Levin R, Munasinghe RL, 1996. Temporal variation in drinking water turbidity and physician diagnosed gastrointestinal infections in Milwaukee. *Am J Public Health*, 86(2): 237-239.

Morris RD, Naumova EN, Griffiths JK, 1998. Did Milwaukee experience waterborne cryptosporidiosis before the large documented outbreak in 1993? *Epidemiol*, 9(3): 264-270.

Murphy PA, Craun GF, 1990. A review of recent epidemiologic studies reporting associations between drinking water disinfection and cancer. In: *Water Chlorination: Chemistry, Environmental Impact and Health Effects* (Jolley et al, editors); Lewis Publishers; Chelsea, Michigan: 361-372.

Ongerth JE, 1989. Giardia cyst concentrations in river water. *J Am Water Works Assoc*, 81(9): 81-86.

Payment P, 1999. Poor efficacy of residual chlorine disinfectant in drinking water to inactivate waterborne pathogens in distribution systems. *Can J Microbiol*, 45: 709-715.

Payment P, Richardson L, Siemiatycki J, Dewar R, Edwardes M, Franco E, 1991. A randomized trial to evaluate the risk of gastrointestinal disease due to the consumption of drinking water meeting current microbiological standards. *Am J Public Health*, 81: 703-708.

Payment P, Siemiatycki J, Richardson L Renaud G, Franco E, Prevost M, 1997. A prospective epidemiological study of gastrointestinal health effects due to the consumption of drinking water. *Internatl J Environ Health*, 1: 5-31.

Perz JF, Ennever FK, Le Blanq SM, 1998. Cryptosporidium in tap water: comparison of predicted risks with observed levels of disease. *Am J Epidemiol*, 147(3) 289-301.

Raucher RS, Dixon AM, Trabka E, Drago JA, 1994. Cost effectiveness of SDWA regulations. *J Am Water Works Assoc*, 88 (8- Aug.): 28-36.

Rose JB, Gerba CP, Jakubowski W, 1991. Survey of potable water supplies for Cryptosporidium and Giardia. *Envir Sci Tech*, 25(8): 1393-1400.

Sax NI, Lewis RJ, 1989. Dangerous Properties of Industrial Materials. Von Nostrand; Reinhold, New York.

Schwartz J, Levin R, Hodge K, 1997. Drinking water turbidity and pediatric hospital use for gastrointestinal illness in Philadelphia. *Epidemiol*, 8: 615-620.

Schwartz J, Levin R, Goldstein R, 2000. Drinking water turbidity and gastrointestinal illness in Philadelphia=s elderly. *J Epidemiol Comm Health*, 54: 45-51.

Singer PC. 1993. Formation and characterization of disinfection by-products. In: *Safety of Water Disinfection: Balancing chemical and microbial risks* (Craun, editor); ILSI (International Life Sciences Institute) Press; Washington, DC: 201-220.

St. Louis ME, 1988. Water-related disease outbreaks, 1985. In: CDC Surveillance Summaries, June 1988. *Morb Mort Wkly Repts*, 37 (No. SS-2): 15-24.

Swan SH, Waller K, Hopkins B, Windham G, Fenster L, Schaefer C, Neutra, RR, 1998. A prospective study of spontaneous abortion: relation to amount and source of drinking water consumed in early pregnancy. *Epidemiology* 9(2): 126-133.

Thomas C, Gibson H, Hill DJ, Mabey M, 1999. Campylobacter epidemiology: an aquatic perspective. *J Appl Microbiol Symposium Supplement*, 85: 168S-177S.

Vena JE, Graham S, Freudenheim J, Marshall J, Zielezny M, Swanson M, Sufrin G, 1993. Drinking water, fluid intake, and bladder cancer in western New York. *Arch Environ Health*, 48(3): 191-198.

Waller K, Swan SH, DeLourenze G, Hopkins B, 1998. Trihalomethanes in drinking water and spontaneous abortion. *Epidemiology* 9(2): 134-140.

Water Quality Association, 1996. Customer Survey, Lisle, Illinois.

Chapter 12

Effects of Land Subsidence in the Greater Houston Area

Devin L. Galloway
U.S. Geological Survey, Sacramento, California
Laura S. Coplin
U.S. Geological Survey, Austin, Texas
Steve E. Ingebritsen
U.S. Geological Survey, Menlo Park, California

12.1 Introduction

The greater Houston area, possibly more than any other metropolitan area in the United States, has been adversely affected by land subsidence (Coplin and Galloway, 1999). Extensive subsidence caused mainly by ground-water extraction has increased the frequency of flooding, caused extensive damage to industrial and transportation infrastructure, motivated major investments in levees, reservoirs, and surface-water distribution facilities, and caused substantial loss of wetland habitat in and around Galveston Bay. Subsidence ultimately limited local ground-water development and drove expensive efforts to store and convey surface water to supplant ground-water pumpage. In the eastern part of the greater Houston region, near the coast, subsidence has been controlled. The area of active subsidence has shifted from the low-lying, tidal areas toward higher elevations inland. Total costs of subsidence damage and mitigation are difficult to assess but likely range in the billions of dollars.

D.E. Agthe et al. (eds.), Managing Urban Water Supply, 187–203.

12.2 Land Subsidence Due to Aquitard Drainage

In alluvial aquifer systems, such as the Texas Gulf Coast aquifer system underlying the greater Houston area shown in Figure 12.1, long-term ground-water-level declines can result in a vast one-time release of "water of compaction" from semiconsolidated silt and clay layers (aquitards), which manifests itself as land subsidence. This release of water results in a largely non-recoverable reduction in the pore volume of the aquitards and thus an overall reduction in the storage capacity of the aquifer system. The extraction of "water of compaction" for economic gain constitutes ground-water mining in the truest sense of the word.

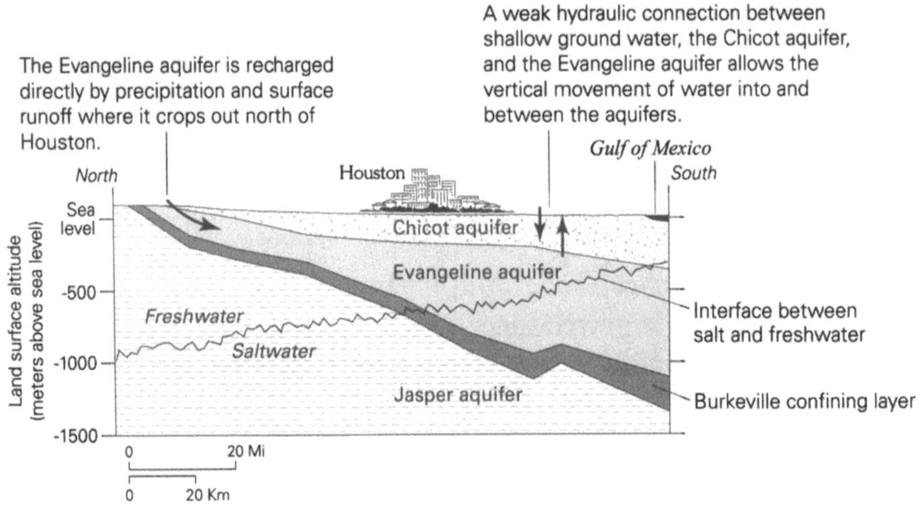

Figure 12.1 Aquifers Underlying Houston, Texas

Most of the ground water pumped in the greater Houston area comes from the Chicot and Evangeline aquifers – part of a vast coastal alluvial-aquifer system that extends throughout the margin of the coastal plain of Texas and Louisiana into Florida (Grubb, 1998). Most supply wells are completed in the upper 300-600 m of the aquifer systems, where freshwater is available. The interface between the freshwater and underlying saltwater slopes landward from the coastline. Saltwater encroachment in both aquifers has been exacerbated by lowered ground-water levels, especially near the coast.

The reduction of fluid pressure in the pores and cracks of aquifer systems, especially in poorly consolidated rocks, is inevitably accompanied by deformation of the aquifer system. Because the granular structure – the so-called "skeleton" – of the aquifer system is not rigid, but more or less compliant, a shift in the balance of support for the overlying material causes the skeleton to deform slightly. Both the aquifers and aquitards that together constitute the aquifer system undergo deformation, but to different degrees. Almost all the permanent subsidence occurs due to the irreversible compression or consolidation of aquitards during the typically slow process of aquitard drainage (Pratt and Johnson, 1926; Tolman and Poland, 1940; Holzer, 1998).

12.2.1 MINOR REVERSIBLE DEFORMATION OCCURS IN ALL AQUIFER SYSTEMS

The relation between changes in ground-water levels and compression of the aquifer system is based on the principle of effective stress first proposed by Karl Terzaghi (1925):

$$\sigma_T = P + \sigma_e \tag{12.1}$$

By this principle, for any arbitrary horizontal plane below the water table, the weight of the overlying rock and water (total stress σ_T) is balanced by the pore-fluid pressure (P) and the intergranular effective stress (σ_e). When the support provided by fluid pressure is reduced, such as when ground-water levels are lowered, the load is transferred to the skeleton of the aquifer system. This results in an increase in intergranular or effective stress, and the skeleton compresses to a degree. Conversely, when the pore-fluid pressure is increased, such as when ground water recharges the aquifer system, support previously provided by the skeleton is transferred to the fluid and the skeleton expands. When the load on the skeleton remains less than any previous maximum load, the fluctuations create only a small elastic deformation of the aquifer system and small displacement of land surface. This fully recoverable deformation occurs in all aquifer systems, resulting in seasonal, reversible displacements in land surface of as much as 6 cm or more in response to the seasonal changes in ground-water pumpage. See Figure 12.2.

12.2.2 INELASTIC COMPACTION IRREVERSIBLY ALTERS THE AQUIFER SYSTEM

The maximum level of past stressing of a skeletal element is termed the preconsolidation stress (Casagrande, 1936). When the load on the aquitard skeleton exceeds the preconsolidation stress, the aquitard will undergo significant, permanent rearrangement, resulting in irreversible compaction. Thus, when long-term pumping lowers ground-water levels and raises stresses on the aquitards beyond the preconsolidation-stress thresholds, the aquitards compact and the land surface subsides permanently as illustrated in Figure 12.2. Pore fluid is "squeezed" out of aquitards into adjacent aquifers. This results in a permanent reduction of pore volume and the storage capacity of the aquitard and a one-time mining of stored ground water. In confined aquifer systems subject to large-scale overdraft, the volume of water derived from irreversible aquitard compaction is essentially equal to the volume of subsidence and typically ranges from 10 to 30 percent of the total volume of water pumped.

Because aquitards are by definition much less permeable than aquifers, the vertical drainage of aquitards into adjacent pumped aquifers may proceed very slowly, and thus lag far behind the changing water levels in adjacent aquifers. For a given head decline in the adjacent aquifers, full compaction of thick aquitards may require decades to approach completion. The theory of hydrodynamic consolidation (Terzaghi, 1925) or aquitard drainage (Holzer, 1998) describes the delay involved in draining aquitards when heads are lowered in adjacent aquifers, as well as the residual compaction that may continue long after drawdowns in the aquifers have essentially stabilized.

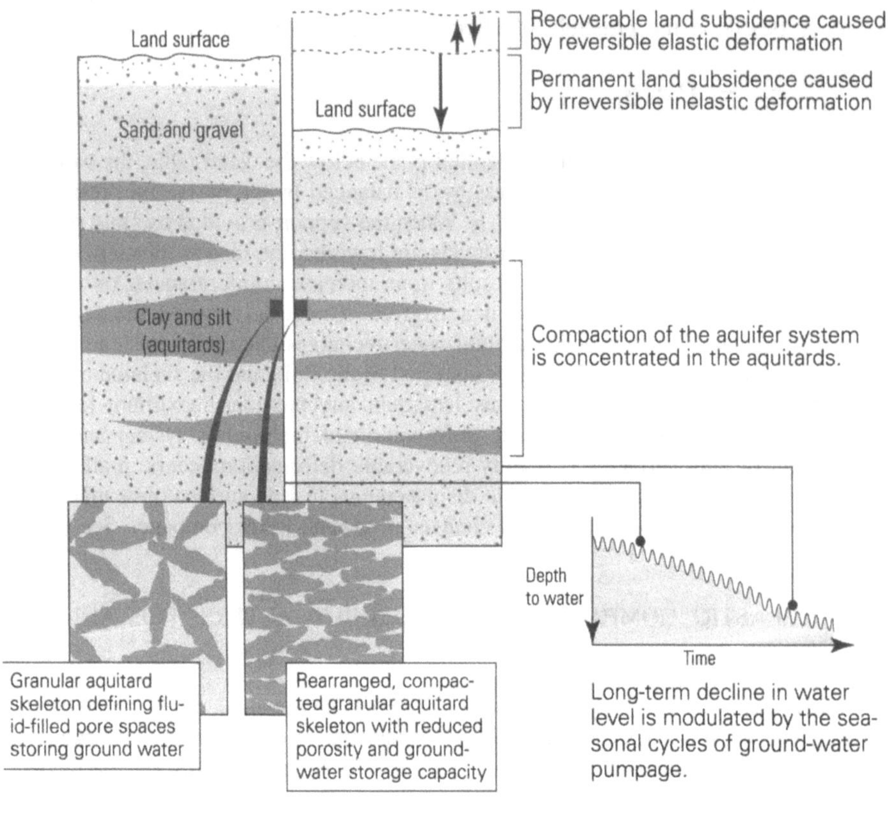

Figure 12.2 Aquitard Compaction

12.3 Regional Setting

Although regional land subsidence is often subtle and difficult to detect, there are localities in and near Houston where the effects are quite evident. In this low-lying coastal environment, as much as 3 m of subsidence has shifted the position of the coastline and altered the distribution of wetlands and aquatic vegetation. In fact, the San Jacinto Battleground State Historical Park, site of the battle that won Texas independence, is now partly submerged. This park, about 20 miles 32 km) east of downtown Houston on the shores of Galveston Bay, commemorates the April 21, 1836 victory of Texans led by Sam Houston over Mexican forces led by Santa Ana. About 0.4 km^2 of the park are now under water due to subsidence, and part of the remaining area must now be protected from the Bay by dikes. At many localities in the Houston area, ground-water pumpage and subsidence have also induced fault movement, leading to visible fracturing, surface offsets, and associated property damage.

The Houston-Galveston Bay area as shown in Figure 12.3 includes a large bay-estuary-lagoon system consisting of the Trinity, Galveston, East and West Bays, which are separated from the Gulf of Mexico by Pelican Island, Galveston Island, and the Bolivar Peninsula. Tidal exchange occurs between the Gulf and bay system through the barrier-island and peninsula complex.

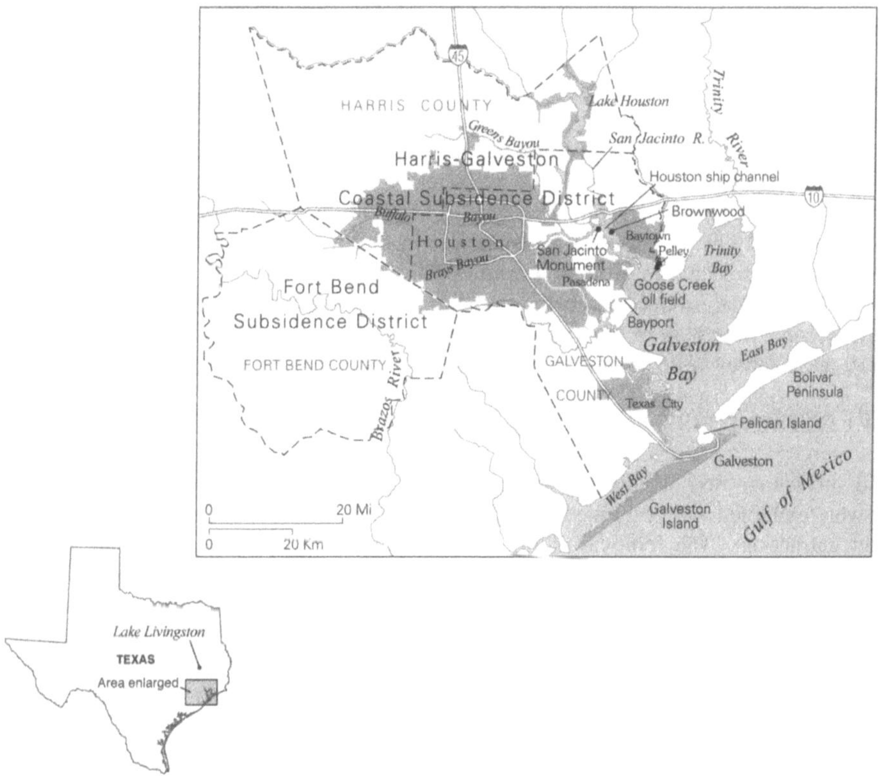

Figure 12.3 Map of the Greater Houston Area.

The Houston climate is subtropical; temperatures range from 7 to 34°C and annual rainfall averages about 1.2 m per year. The coastal plain slopes gently toward the Gulf at a gradient of about 0.0003. Two major rivers, the Trinity and San Jacinto, and many smaller ones traverse the plain before discharging into estuarine areas of the bay system. The warm waters of the Gulf of Mexico are conducive to hurricanes and tropical storms, and the Texas coast is subject to a hurricane or tropical storm about once every 2 years (McGowen and others, 1977). The flat-lying region is particularly prone to both coastal and riverine flooding, and the rivers, their reservoirs, and an extensive system of bayous and manmade canals are managed as part of an extensive flood-control system.

12.4 Houston Grows and Subsides

Since 1897, when the population was about 25,000, the Houston area has experienced rapid growth, spurred by the discovery of oil and establishment of the Port of Houston. In 1907 the first successful oil well was drilled, marking the beginning of the petrochemical industry that provided the economic base on which the Houston area was built and still stands. In 1914 Houston became a deep-water port when the U.S. Army Corps of Engineers completed dredging the Houston Ship Channel across Galveston Bay, up the lower reaches of the San Jacinto River, and along Buffalo Bayou to Houston. Easy access to the Gulf via the ship channel, and the discovery of additional oil fields, triggered major industrial development along the ship channel in Baytown, La Porte, Pasadena, Texas City, and Houston. The region and industry have continued to grow, and by 1999 the Houston-Galveston area had a population of about 3 million people that is projected to grow to 4.5 million by the year 2010. Nearly one-half of all U.S. petrochemical production occurs in the greater Houston area. The Port of Houston is the second largest port (by tonnage shipment) in the nation, eighth largest in the world, and handles more commodities for Mexico than all Mexican ports combined.

12.4.1 INITIAL SUBSIDENCE DUE TO OIL AND GAS EXTRACTION

Land subsidence first occurred in the early 1900s in areas where ground water, oil, and gas were extracted and continued throughout the 20th century due primarily to ground-water extraction. The removal of oil and gas along with the attendant brine, ground water and sand first caused localized subsidence above oil fields such as Goose Creek (Figure 12.3). The subsidence of Goose Creek oil field was the first subsidence due to subsurface-fluid withdrawal to be described in the scientific literature (Pratt and Johnson, 1926). Subsidence, accompanied by fissuring caused the center of the Goose Creek field to be submerged beneath Galveston Bay. The State of Texas owns submerged land, and claimed the field after it became submerged. A subsequent legal dispute about the status of the land caused Texas courts to formally recognize the process of subsidence due to subsurface fluid withdrawal.

12.4.2 SUBSIDENCE REFLECTS EVOLVING PATTERNS OF FLUID EXTRACTION

Subsequent patterns of subsidence in the Houston area, shown in Figure 12.4, closely follow the temporal and spatial patterns of activity, shifting inland from coastal oil and gas development to ground-water extraction for industrial and municipal supplies. Near Texas City, the withdrawal of ground water for public supply and industry caused more than 0.5 m of subsidence between 1906 and 1943. This period also marked the development of ground-water resources as the sole water supply for industries and communities around the Ship Channel, including Houston. By 1937 ground-water levels were falling in a growing set of gradually coalescing cones of depression centered on the areas of heavy use. By 1943 subsidence had begun to affect a large part of the Houston area, although the amounts were generally less than 0.3 m (Figure 12.4a).

A period of rapid growth in the development of ground-water resources was driven by the expansion of the petrochemical industry and other allied industries in the late 1930s through the late 1970s. By the mid-1970s, 2 m or more of subsidence had occurred throughout an area along the Houston Ship Channel between Bayport and Houston as a result of declining ground-water levels associated with the rapid industrial expansion (Figure 12.4b). By 1979, as much as 3 m of subsidence had occurred in Baytown and more than 8,000 km^2 had subsided at least 30 cm.

12.5 Subsidence Exacerbates Flooding

During the 1970s, the consequences of subsidence problems in Houston were becoming more widely recognized. In coastal areas, subsidence had increased the amount of land subject to the threat of tidal inundation. The fate of the Brownwood subdivision of Baytown (Figure 12.3) afforded a particularly dramatic example of the dangers of coastal subsidence. Brownwood was constructed, beginning in 1938, as an upper-income subdivision on wooded lots along Galveston Bay (Holzschuh, 1991). At that time the area was generally 3 m or less above sea level. By 1978 more than 2.4 m of subsidence had occurred. The fate of Brownwood was reported by Gabrysch (1983).

> The subdivision is ... a community of about 500 single-unit family houses. Because of subsidence, a perimeter road was elevated in 1974 to allow ingress and egress during periods of normal high tide ... and to provide some protection during unusual high tide. Pumps were installed to remove excess rainfall from inside the leveed area. Because of subsidence after the roadway was elevated, [large] tides ...will cause flow over the road. The United States Army Corps of Engineers studied methods to protect the subdivision from flooding. The cost of a levee system was estimated to be about $70 million. In 1974, the Army Corps estimated that it would cost about $16 million to purchase 442 homes, relocate 1,550 people, and convert [much] of the peninsula into a park. This proposed solution was approved by the

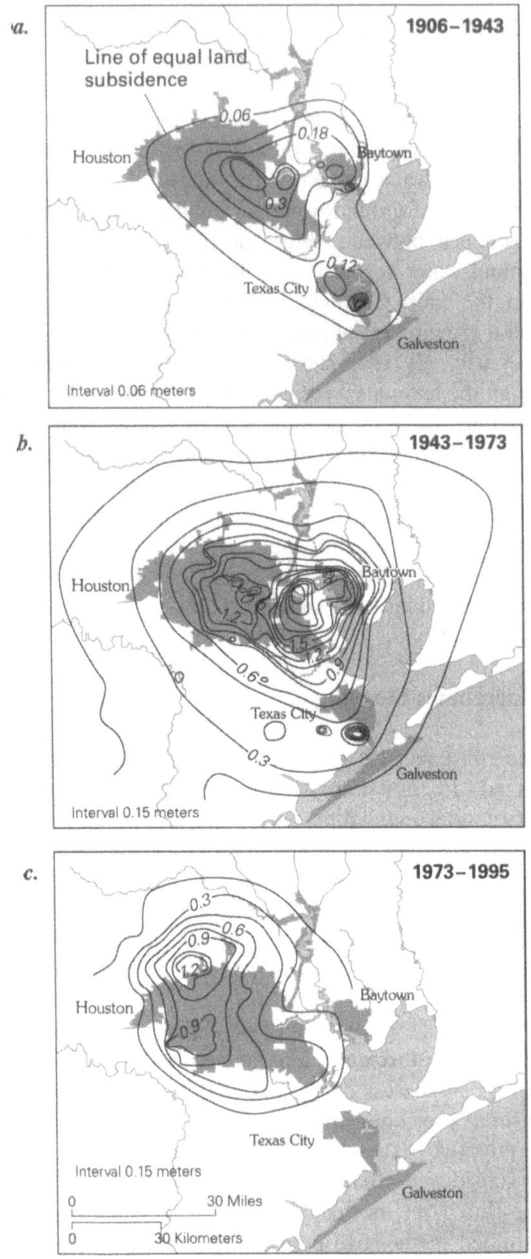

Figure 12.4 Evolving Subsidence (a) 1906-1943; (b) 1943-1973; (c) 1973-1995
Source:Harris-Galveston Coastal Subsidence District

> Congress of the United States which provided the necessary funding. However, the project required that a local sponsor (the City of Baytown) should ... provide 20 per cent of the funds ... By the time the first election to fund the project was held on 23 July 1979, the cost estimate had increased to $37.6 million, of which the local share was $7.6 million. The proposal was defeated, and two days later 12 inches (30.5cm) of rain fell on Brownwood causing the flooding of 187 homes. Another bond election was held on 9 January 1980 and again the proposal was defeated. Accepting the residents' decision, Baytown officials began planning the sale of $3.5 million worth of bonds to finance the first stage of a fifteen-year, $6.5-million program to upgrade utilities in the subdivision. Meanwhile, those who own the houses generally also owe mortgages and cannot afford to purchase other homes. Although they continue to live in the subdivision many have to evacuate their homes about three times each year.

The year that article was published, Hurricane Alicia struck a final blow to Brownwood. All homes in the subdivision were abandoned and eventually removed. Today, the site of Brownwood is the Baytown Nature Center. Otters, alligators, and a wide variety of waterfowl are frequently observed.

In addition to its dramatic effects along the coast, subsidence exposes inland areas to increased risks of flooding and erosion by altering natural and engineered drainageways that depend on gravity-driven flow of storm runoff and sewerage. Differential subsidence can either reduce or enhance preexisting gradients. Gradient reductions decrease the rate of drainage and thereby increase the chance of flooding by storm-water runoff. Gradient reversals can result in ponding or backflow of sewage and stormwater.

12.6 Texas and Houston Act to Arrest Subsidence

The increased incidence of flooding in coastal areas led to growing public awareness of subsidence and its costs. In the late 1960s and early 1970s groups of citizens began to work for a reduction in ground-water use. As early as the 1940s, upstream reservoirs and canals had allowed the first deliveries of surface water to coastal areas east and south of Houston. Further imports of surface water from Lake Houston (Figure 12.3), a reservoir on the San Jacinto River completed in 1954, locally and temporarily halted water-level declines, but proved insufficient to keep pace with the growing demand for water. Ground water remained the primary source in most of the coastal area until the mid-1970s.

12.6.1 SPECIAL DISTRICTS CHARTERED TO CONTROL SUBSIDENCE

In 1975 the Texas Legislature, prompted by concerned citizens and business leaders, passed a law creating the Harris-Galveston Coastal Subsidence District, the first district of its kind in the United States (http://hgsubsidence.org). The unprecedented subsidence district was authorized as a regulatory agency, with the power to restrict ground-water withdrawal by annually issuing or denying permits for large-diameter wells, but was forbidden to own property such as water-supply and conveyance facilities.

The Subsidence District, in 1976, focused initially on the critical situation in the coastal areas. It prompted industry along the Houston Ship Channel to convert from ground water to surface water, using the supply from the recently completed Lake Livingston reservoir on the Trinity River. The water supplied from Lake Livingston was sufficient to significantly reduce ground-water use in the Baytown-Pasadena area, resulting in a dramatic slowing of subsidence and ultimately to recovery of ground-water levels over a large area.

Although subsidence in the southeastern part of the greater Houston region, near the bay system, has been effectively controlled by conversion from ground water to imported surface water (Figure 12.4c), subsidence is accelerating to the north and west, where ground-water use has increased. As a result, the area of active subsidence has shifted from the low-lying, tide-affected areas toward higher elevations inland.

A devastating flood in 1984 on Brays Bayou, a major watershed in southwest Houston, renewed concern about the effects of subsidence in inland areas. It was recognized that flood control and subsidence control should be coordinated to minimize flood damages. During the 1989 legislative session, the Fort Bend Subsidence District (Figure 12.3) was created to manage and control subsidence in Fort Bend County, which includes most of the inland subsidence-affected area.

12.6.2 THE ONGOING CONVERSION FROM GROUND WATER TO SURFACE WATER

Since the late 1970s subsidence has largely been arrested along the ship channel and in the coastal Baytown-Texas City area due to a reduction in ground-water pumpage (Figure 12.4c). By 1996, total annual ground-water pumpage in the Houston area had declined to only 60 percent of peak amounts pumped during the mid-1970s (Figure 12.5); within the jurisdiction of the Harris-Galveston Coastal Subsidence District, ground-water pumpage constituted only 25 percent of peak amounts. However, as subsidence in the coastal area stabilized, subsidence inland – north and west of Houston – accelerated. In this region ground-water levels declined more than 30 m in the Evangeline aquifer between 1977 and 1997, and more than 1 m of subsidence was measured immediately northwest of Houston between 1973 and 1995 (Figure 12.4c).

In 1992 the Harris-Galveston Coastal Subsidence District adopted a regulatory action plan to reduce ground-water pumpage by 80 percent no later than the year 2020. Due to the high cost of constructing distribution lines westward across the metropolitan area, the plan was to be implemented in phases, allowing time to design, finance, and construct surface-water importation facilities. The Harris-Galveston and Fort Bend Subsidence Districts intend to cooperate to ensure coordinated planning of the conversion from ground water to surface water. Some controversy attends efforts to achieve conversion on the north and west sides of Houston, mainly because the imported surface water is expected to cost about three times as much as the ground water that is

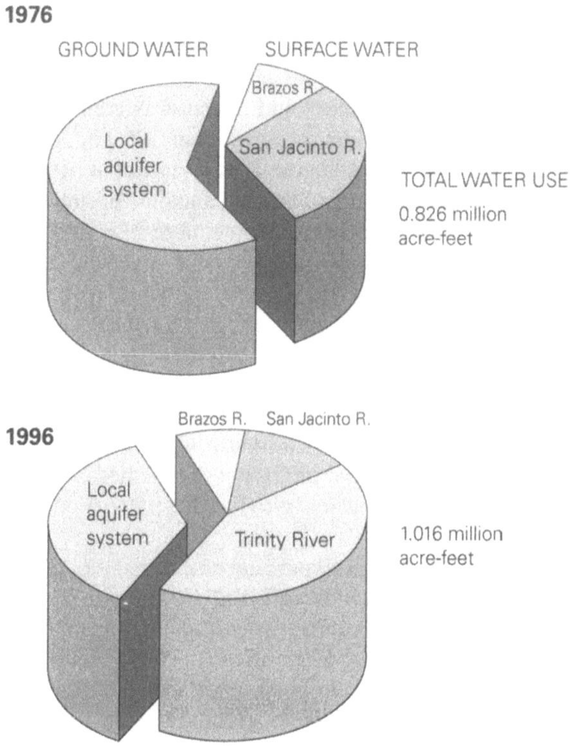

Figure 12.5 Houston Water Use by Source: 1976 and 1996

currently used. Various local municipalities are contesting the timing and apportioning of costs (*Houston Chronicle*, 1997). Efforts to build public support are hampered by the fact that the subsidence damage – while real – is simply not as obvious as it was near the coast.

Given the continuing rapid growth of Houston, there is also a long-term concern about securing sufficient surface-water supplies. State and local governments are seeking to ensure that there will be enough water for the expected future population. The primary strategies aim to promote water conservation and acquire supplies from East Texas reservoirs. In addition to the concerns of East Texas communities about water being exported to Houston, such water transfers have ecological effects on the coast and on the waterways through which the water is moved.

12.7 Wetlands Lost to Subsidence

Galveston Bay is arguably one of the most significant bay ecosystems in the Nation. The estuary is Texas' leading bay fishery and supports vibrant recreation and tourism industries. Mainly because of subsidence, more than 100 km^2 of emergent wetlands have been converted to open water and barren flats (White et al., 1993). Subsidence has also contributed to a significant loss of submerged aquatic vegetation (mostly sea grass) since the 1950s. Some bay shorelines have become more susceptible to erosion by wave action due to loss of fringing wetlands. At the same time, the reduction in sediment inflows to the bay system, resulting from construction of reservoirs along tributary rivers, slows the natural rebuilding of shorelines.

Because of the combined and interrelated effects of relative sea-level rise, loss of wetlands, and reduced sediment supply, many parts of the shoreline have suffered rapid erosion (Paine and Morton, 1986). As the water level rises, marsh along the shoreline is drowned. When residential, commercial or industrial development is located near the shoreline, the potential for the landward migration of marshes is eliminated. The result is a reduction in wetland habitats, which provide the foundation for commercial and recreational fisheries.

The most extensive changes in wetlands have occurred along the lower reaches of the San Jacinto River near its confluence with Buffalo Bayou. This area had subsided by 1 m or more by 1979, resulting in submergence and changes in wetland environments that progressed inland along the axis of the stream valley (Figure 12.6). Open water displaced riverine woodlands and swamps. Trends along the lower reaches of other rivers, bayous, and creeks have been similar, resulting in an increase in the extent of open water, loss of inland marshes and woodlands and, in some areas such as the Brownwood site, the development of new marshes inland from the encroaching waters.

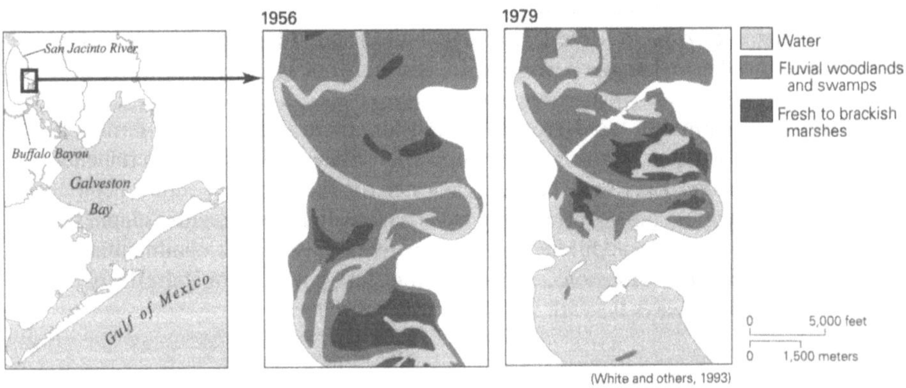

Figure 12.6 Wetlands Lost to Inundation, Lower Reaches of the San Jacinto River (after White and others, 1993).

The health and productivity of the bay ecosystem depends on the mix of river and bay water and nutrients, as well as the presence of key habitats like salt marshes (Galveston Bay National Estuary Program, 1995). Many species of fish, wildlife, aquatic plants and shellfish in Galveston Bay depend on freshwater inflows for survival. The estuary is adapted to highly variable inflows of freshwater. For instance, oysters prefer somewhat salty water but need occasional surges of freshwater. The volume, timing and quality of freshwater inflows to the estuary are key factors. Surface impoundments such as Lake Livingston and Lake Houston, the construction of which was motivated largely by efforts to mitigate land subsidence, have changed the natural freshwater inputs to the bay system: high flows are lower, low flows are higher, and peak flows are delayed by about 1 month. These changes affect the natural processes that evolved with the rhythm of the unregulated streams and rivers.

12.8 Subsidence Activates Faults

Many faults exist in the Houston area, both regional-scale "down-to-the-coast" faults that represent a slow sliding and compaction of alluvium toward the Gulf over geologic time, and local structures associated with oil fields (Holzshuh, 1991). Since the 1920s, active faults have offset the land surface and damaged buildings and highways in the metropolitan area (Holzer and Gabrysch, 1987). The scarps typically grow by seismic creep at rates of as much as 2.5 cm per year (Holzer, 1984). Monitoring of fault creep, water levels, and land subsidence has demonstrated a clear cause-and-effect relation. The fault movement is caused by water-level decline and associated subsidence. When water-level recovery began in the eastern part of the Houston area in the 1970s, fault creep stopped or slowed in the area of water-level recovery but continued in the area of ongoing water-level decline (Figure 12.7).

12.9 The Direct and Indirect Costs of Subsidence

The low elevation, proximity to bays and the Gulf of Mexico, dense population, and large capital investment make it likely that the Houston-Galveston area has been more significantly affected by subsidence than any other metropolitan area in the United States. The actual economic cost of subsidence is hard to quantify, and most published estimates are necessarily vague. For example, Gabrysch (1983) stated that "many millions of dollars" have been spent reclaiming land submerged by tidal water, elevating structures such as buildings, wharves, and roadways, and constructing levees to protect against tidal inundation; further, "millions of dollars" are spent on repairing damage due to fault movement. One conservative estimate for the period 1969 to 1974 placed the average annual cost to property owners at more than $31 million in 1975 dollars (Jones, 1976) or about $100 million in 2000 dollars. The costs of such subsidence-related phenomena as the loss of wetlands are even more difficult to assess than property losses. Although some estimates might be made based on the changing value of commercial and recreational fisheries, it would be difficult to distinguish the influence of subsidence

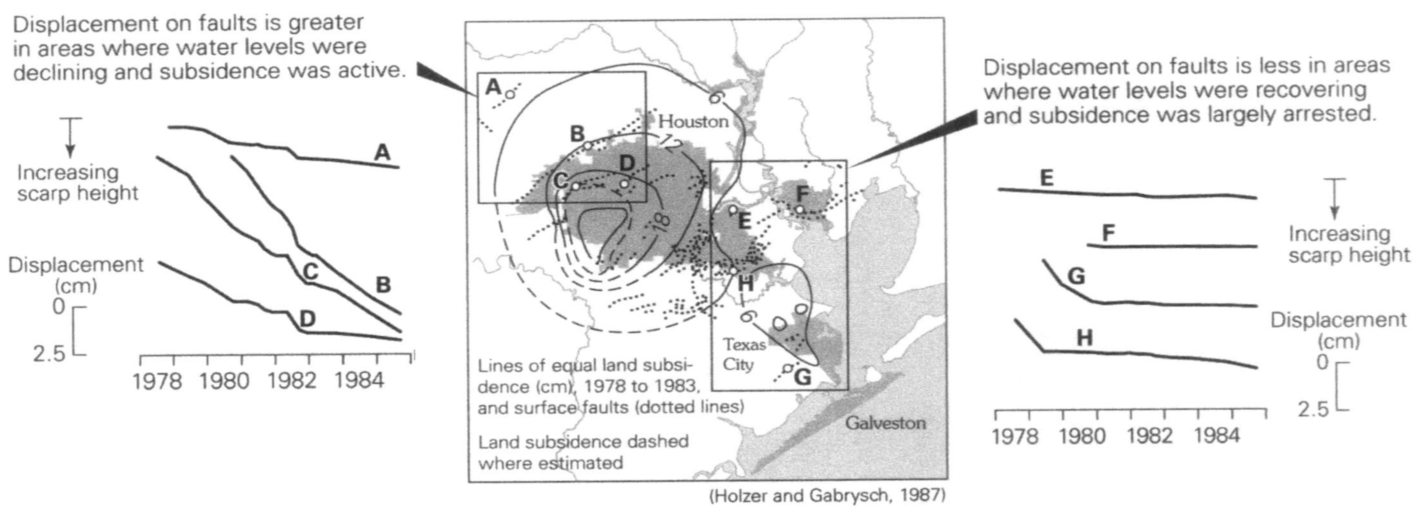

Figure 12.7 **Vertical Displacements at 8 Selected Fault-Monitoring Sites in the Houston area (after Holzer and Gabrysch, 1987).**

from that of other factors. Similarly, some fraction of the ongoing cost of flood prevention and flood damage repair could be fairly attributed to subsidence.

The most definitive published subsidence damage estimates have to do with the cost of relocating dock facilities, constructing hurricane levees and rectifying drainage problems at refineries along the Houston Ship Channel. For two refineries alone, the estimated total cost was $120 million in 1976 dollars (Holzschuh, 1991), or about $360 million in year 2000 dollars. If these estimates are correct, subsidence-related damage to industrial infrastructure alone may run into the billions of dollars. As was the case in other subsidence-affected regions of the United States such as the Santa Clara (Poland and Ireland, 1988; Ingebritsen and Jones, 1999) and San Joaquin Valleys (Poland and others, 1975; Galloway and Riley, 1999) and south-central Arizona (Carpenter, 1999), subsidence ultimately inhibited local ground-water development and helped to drive expensive efforts to store and convey surface water that would supplant ground water.

The ongoing patterns of subsidence in the Houston area, shown in Figure 12.8, are carefully monitored. Compaction of subsurface material is measured continuously using 13 borehole extensometers (wells equipped with compaction monitors) at 11 sites throughout the region (http://hgsubsidence.org).

Synthetic aperture radar imagery for the western Houston area including Jersey Village

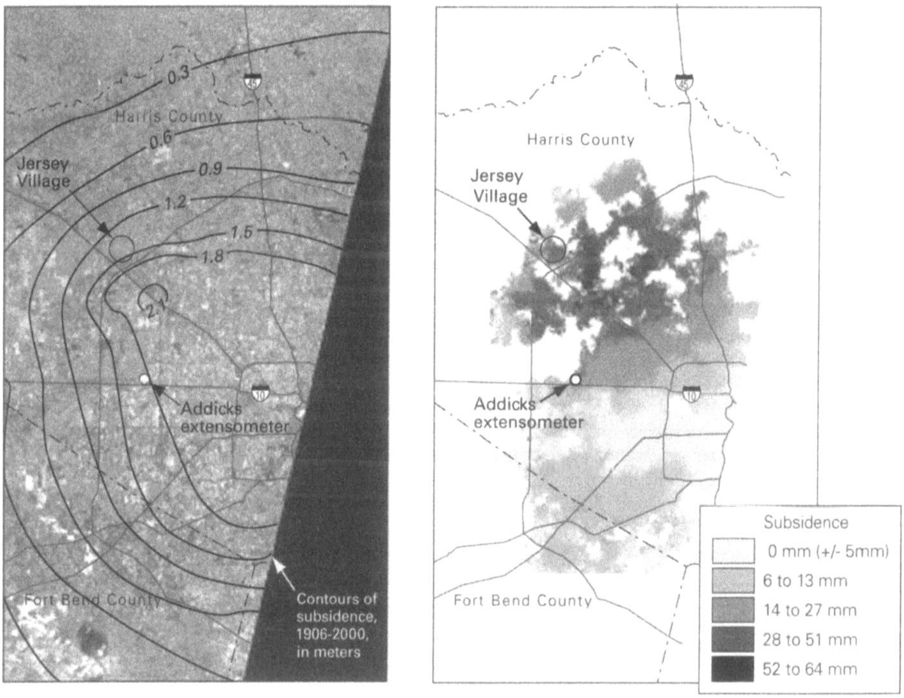

a. *b.*

Figure 12.8 SAR Imagery Showing Subsidence

The elevations of three extensometers are continuously monitored using the Global Positioning System (GPS), under the direction of the National Geodetic Survey. Piezometers completed to different depths are used to simultaneously monitor water levels (http://tx.usgs.gov). A network of 82 bench marks distributed throughout the two subsidence districts was installed in 1987 to determine elevation changes using GPS, and was resurveyed in 1995 and 2000. The GPS approach is expected to be more cost-effective than constructing additional extensometers or surveying benchmarks using traditional leveling techniques, and satellite-radar interferometry shows promise of further reducing costs while increasing resolution.

Houston's continuing rapid growth means that subsidence must continue to be vigilantly monitored and managed. However, the region is well-positioned to deal with future problems for several reasons: a raised public consciousness; the existence of well-established subsidence districts with appropriate regulatory authority; and the knowledge base provided by abundant historical data and ongoing monitoring.

Acknowledgments

We thank Bob Gabrysch and Tom Holzer for helpful review comments and the Harris-Galveston Coastal Subsidence District, the Fort Bend Subsidence District, the National Geodetic Survey, and Eurimage Corporation for information and cooperation.

References

Carpenter, M.C., 1999. South-central Arizona – Earth fissures and subsidence complicate development of desert water resources in Galloway, D.L., Jones, D.R., and Ingebritsen, S.E., 1999, *Land subsidence in the United States*: U.S. Geological Survey Circular 1182: 65-78.

Casagrande, A., 1936. The determination of the pre-consolidation load and its practical applications, *Proceedings of the First International Conference on Soil Mechanics*, 3: 60-64.

Coplin, L.S., and Galloway, D.L., 1999. Houston-Galveston, Texas – Managing coastal subsidence in Galloway, D.L., Jones, D.R., and Ingebritsen, S.E., 1999. *Land subsidence in the United States*: U.S. Geological Survey Circular 1182: 35-48.

Gabrysch, R.K., 1983. The impact of land-surface subsidence in managing our fresh-water resources: Paris, United Nations Educational, Scientific, and Cultural Organization [UNESCO], *Impact of Science on Society*, 1: 117-123.

Galloway, D.L., Jones, D.R., and Ingebritsen, S.E., 1999. *Land subsidence in the United States*: U.S. Geological Survey Circular 1182, 175 p.

Galloway, D.L., and Riley, F.S., 1999. San Joaquin Valley, California – Largest human alteration of the Earth's surface in Galloway, D.L., Jones, D.R., and Ingebritsen, S.E., 1999. *Land subsidence in the United States*: U.S. Geological Survey Circular 1182: 23-34.

Galveston Bay National Estuary Program, 1995. *The Galveston Bay plan – The comprehensive land management plan for the Galveston Bay ecosystem*, Oct. 18, 1994: Galveston Bay National Estuary Program Publication GBNEP-49, 457 p.

Grubb, H.F., 1998. *Summary of the hydrology of the regional aquifer systems, Gulf Coastal Plain, south-central United States*: U.S. Geological Survey Professional Paper 1416-A, 61 p.

Holzer, T.L., 1984, Ground failure induced by ground-water withdrawal from unconsolidated sediment in Holzer, T.L., ed., *Man-induced land subsidence: Geological Society of America Reviews in Economic Geology*, v. 6, p. 67-105.

Holzer, T.L., 1998, History of the aquitard-drainage model in Borchers, J.W., ed., *Land subsidence case studies and current research: Proceedings of the Dr. Joseph F. Poland Symposium on Land Subsidence*: Association of Engineering Geologists Special Publication 8: 7-12.

Holzer, T.L., and Gabrysch, R.K., 1987, Effect of water-level recoveries on fault creep, Houston, Texas: *Ground Water*, 25: 392-397.

Holzschuh, J.C., 1991. *Land subsidence in Houston, Texas U.S.A.*: Houston, Field-trip guidebook for the Fourth International Symposium on Land Subsidence, May 12-17, 22 p.

Houston Chronicle, 27 August 1997, "That sinking feeling hits northwest Houston"

Ingebritsen, S.E., and Jones, D.R., 1999. Santa Clara Valley, California – A case of arrested subsidence in Galloway, D.L., Jones, D.R., and Ingebritsen, S.E., 1999. *Land subsidence in the United States*: U.S. Geological Survey Circular 1182: 15-22.

Jones, L.L., 1976. External costs of surface subsidence – Upper Galveston Bay, Texas in *Proceedings of the 2nd International Symposium on Land Subsidence*, Anaheim, California, December: International Association of Hydrological Sciences Publication 121: 617-627.

McGowen, J.M., Garner, L.E., and Wilkinson, B.M., 1977. *The Gulf shoreline of Texas—Processes, characteristics, and factors in use*: University of Texas, Bureau of Economic Geology, Geological Circular 75-6, 43 p.

Paine, J.G., and Morton, R.A., 1986. *Historical shoreline changes in Trinity, Galveston, West, and East Bays, Texas Gulf Coast*: University of Texas, Bureau of Economic Geology, Circular 86-3, 58 p.

Poland, J.F., and Ireland, R.L., 1988. *Land subsidence in the Santa Clara Valley, California, as of 1982*: U.S. Geological Survey Professional Paper 497-F, 61 p.

Poland, J.F., Lofgren, B.E., Ireland, R.L., and Pugh, R.G., 1975. *Land subsidence in the San Joaquin Valley, California as of 1972*: U.S. Geological Survey Professional Paper 437-H, 78 p.

Pratt, W.E., and Johnson, D.W., 1926. Local subsidence of the Goose Creek oil field: *Journal of Geology*, v. 34: 577-590.

Terzaghi, K., 1925. Principles of soil mechanics, IV—Settlement and consolidation of clay: *Engineering News-Record*, 95: 874-878.

Tolman, C.F., and Poland, J.F., 1940. Ground-water infiltration and ground-surface recession in Santa Clara Valley, Santa Clara County, California: *Transactions American Geophysical Union*, 21: 23-34.

White, W.A., Tremblay, T.A., Wermund, E.G., Jr., and Handley, L.R., 1993. *Trends and status of wetland and aquatic habitats in the Galveston Bay system, Texas*: Galveston Bay National Estuary Program Publication GBNEP-31, 225 p.

Chapter 13

Solving Groundwater Overdraft in Arizona Urban Areas

Donald E. Agthe
Economic Consultant
R. Bruce Billings
University of Arizona

13.1 Introduction

Arizona's rapid economic and population growth since 1950 has relied heavily on mined groundwater. With limited surface water because of its generally arid climate, many parts of the State have experienced considerable groundwater overdraft. In 1980 the Arizona State Legislature passed the Groundwater Management Act (GMA) which was intended to limit ground water pumping for both agricultural and urban use in order to reduce the adverse effects associated with aquifer depletion. The GMA created administrative units called Active Management Areas (AMA) to carry out this reduction in groundwater use.

The regions defined as AMAs are based on identified hydrologic areas that are somewhat similar to those defined by existing county political boundaries. Phoenix, Tucson, and Prescott, are the three urban areas considered in this discussion for which AMAs were created. There is also an AMA for Nogales, which is on the Mexican border. The part of Nogales within Mexico is much larger than the portion in Arizona, and presents a number of special challenges which are not considered here.

While the geography, climates, economies, and populations of Phoenix, Tucson, and Prescott are very different, these cities are similar in that they are rapidly growing in

D.E. Agthe et al. (eds.), *Managing Urban Water Supply*, 205–218.

population and are overdrafting their local groundwater supplies. Phoenix, with a metropolitan population in excess of 3 million, houses the state capital. It is at an elevation of 1100 feet and receives the drainage from 2/3 of the state. Hence this area has made extensive use of river water for agricultural irrigation. Irrigation water delivery has continued in those parts of the area that have been converted from farm land. This limited geographic distribution of previously agricultural water is partially responsible for the fact that about one-third of the urban water consumed in Phoenix is from groundwater sources.

Tucson, with a metropolitan population of a little over 800,000 residents, is 100 miles (161 km) south-east of Phoenix at an elevation of 2400 feet (679 m). It receives runoff from nearby mountains, but these flows are ephemeral and do not provide a reliable source of surface water. Historically, Tucson has depended entirely on ground water. With rapidly increasing population, combined with copper mining and agricultural development, the area has experienced rapid aquifer depletion.

Prescott, a small city of a little over 30,000 people located 96 miles (154 km) northwest of Phoenix at an elevation of 5387 feet (1525 m) above sea level, is dependent upon ground water imported from nearby valley area wells. It receives very little runoff, being higher in elevation than most of its surrounding area. However it benefits from a cooler climate and higher precipitation, as shown in Table 13.1 which compares the climates of these cities.

Table 13.1 Arizona Temperature and Precipitation

	Phoenix	Prescott	Tucson
Summer Mean Temperature	91^0F (33^0C)	70^0F (21^0C)	86^0F (30^0C)
Winter Mean Temperature	55^0F (13^0C)	37^0F (3^0C)	54^0F (12^0C)
Annual Precipitation	7.8 in (20cm)	19.5 (50cm)	12.2 (31cm)

Source: U.S Weather Service

The Central Arizona Project (CAP) brings water from the Colorado River to Phoenix and Tucson as well as to agricultural users and several smaller cities near the canal route. Tucson is at the south-east end of the 336 mile (541 km) canal system. Although the CAP was originally intended to supply irrigation water for agricultural expansion, its primary purpose now is to reduce over-drafting of the state's aquifers by replacing mined groundwater with river water. Thus the CAP is important to the achievement of the GMAs groundwater use reduction objectives. Despite the CAP's importance, Arizona's water users are expected to repay no more than one-half of the project's cost over the 50-year repayment period. Prescott and Nogales do not have access to CAP water because of their higher elevations and distances from the canal.

13.2 The Arizona Groundwater Management Act and the AMAs

The GMA's primary stated goals are to control severe overdraft of several aquifers that serve cities and agriculture in central and southern Arizona, provide a means for allocating scarce groundwater resources, and develop additional water supplies (Carpenter, 1999). The GMA is intended to protect critical aquifers from future depletion rather than restore groundwater levels that existed 40 or 50 years ago before over-drafting became common. Many observers allege that an unstated but important goal of the Arizona GMA legislation was to persuade the Carter administration (1976) that Arizona was addressing its water problems so that the CAP would be built using $4 billion of federal government funds.

The GMA recognized groundwater as a public resource rather than the private resource it had been in the past. This shift was necessary because aquifers are common pool resources. Each well owner has an incentive to increase water use without regard to other users as long as the value provided by additional water is greater than the cost of pumping. This condition leads to overdraft of the aquifer.

The primary goal of the GMA and the AMAs that it created is to achieve safe yield by the year 2025, so that groundwater withdrawals will not continue to exceed recharge. The AMAs have worked toward their goals by implementing regulations on water use, providing educational programs, prohibiting extension of irrigation to new areas, and supplementing local water supplies with imported water.

13.3 The Safe Yield Concept and Problems of Measurement

The concept of *safe-yield* is central to the goals of the GMA. Safe-yield, as defined by the GMA, is the amount of water that can be withdrawn while maintaining a long-term balance between annual groundwater withdrawals and natural and artificial recharge (A.R.S. 45-561.13). One problem with this definition is that there are few if any accurate measures of natural recharge. Safe yield is also difficult to forecast since annual recharge and therefore allowable volume of groundwater that can be withdrawn are highly variable. Thus, allowable water use is not known when agricultural crop plans for the next season are being put into place. This makes conditional transfers from agriculture to urban use difficult. The long-term balance requirement as stated in the GMA could be construed to allow for storage in years of abundant water supply and net extraction in years of drought. Excess withdrawals during droughts would be replaced by natural recharge and/or increased artificial (direct) recharge when supplies were larger.

Direct (artificial) recharge, water pumped or allowed to seep into the ground from effluent or from out-of-basin sources, can be measured with reasonable accuracy, although its later withdrawal is somewhat problematic since water stored in natural aquifers tends to migrate down gradient. The availability of surplus water which might be recharged will vary over time with annual precipitation as well as political, legal and economic developments. For example, CAP water may only be available for recharge in years when Arizona's normal allocation is not being fully used or when there are larger than normal flows in the Colorado River. The options for recharging effluent are determined by public acceptance, environmental regulations, and other factors.

Incidental (natural) recharge varies because of wet and dry years, such as those associated with El Nino (extreme wet events) and La Nina (extreme dry events) which recur in cycles of unpredictable length and duration. Tree ring research has demonstrated that Arizona has experienced long drought cycles in the past which reduce incidental recharge (as well as surface water supplies) for multiple years or even decades. Runoff from melting snow packs in mountain ranges is somewhat unpredictable as to timing and amount. The capacity of recharge zones depends on soil type and subterranean geography as well as changes in surface land use, stream flow alterations, and the overdraft itself. Parking lots, streets, and other surface structures inhibit recharge, especially when storm drains remove water from what were originally porous arroyos before urbanization. Urban recharge is often of reduced quality due to pollutants.

Safe yield applied to an entire AMA does not mean that all areas of the AMA will be in safe yield because areas of high demand for groundwater and high recharge may not coincide. For example, the northern part of the Phoenix AMA and the southern and central parts of the Tucson AMA may experience severe overdraft while areas downstream of their respective sewage treatment plants may have rising water tables.

13.4 Costs of Overdraft

The costs of overdraft of ground water include loss of supply, land subsidence, aquifer compaction, and water quality degradation (Phoenix, 1999, p. 13-6). As the water table falls, streams flows are reduced or eliminated. Riparian habitat for animals, birds, and plants is endangered and the local ecology modified. Overdraft of groundwater lowers depths in wells as the water table falls, leading to increased costs of drilling and pumping. It also leads to a reduction in water quality in areas of heavy mineralization of soils. Soil compaction occurs, reducing the ability of the aquifer to hold water and slowing subsurface flows. Fissuring and subsidence are a long run result of groundwater overdraft. The capacity of an aquifer is likely to be permanently reduced by subsidence, since pores in the soil previously containing moisture are compacted. Subsidence can also cause cracked foundations and plaster in structures, breaks in canals and underground utility lines, broken and sunken pavement in roadways, reduced productivity in existing wells, and greater flood damage from lowered elevations of real property. In the Phoenix AMA for example, subsidence of approximately six inches (15cm) in Scottsdale damaged a number of sewer lines whose gravity flow system was disrupted. Subsidence also resulted in $16 million in repairs to Dysart Dam and $22 million for repairs at Luke Air Force Base (Gelt, 1992). The central area of Tucson has sunk several inches with few visible impacts so far (Carpenter, 1999).

13.5 Costs of Achieving Safe Yield

The attempt to achieve safe-yield has incurred various costs to society, including investments to bring water from distant sources, shifts to lower quality water, and government administrative expansion for aquifer management. Water conservation requirements may lower crop yields, reduce industrial productivity, and change production methods, bringing lower earnings to entrepreneurs. Business and agriculture face an additional set of regulations to be complied with, including new record keeping and monitoring activities.

13.6 General Analysis of the AMAs

Table 13.2 shows the 1985 base line and the 1998 data on ground water used in each AMA by major sector. Each of the three AMAs has achieved a decrease in the use of groundwater as a percent of total water used since 1985. The Phoenix and Prescott AMAs also reduced the absolute amount of groundwater used. In contrast, the Tucson AMA recorded an increase in groundwater use over this period because of a failure to utilize CAP water. Most of the large reductions in total groundwater used in the urban AMAs came in the agricultural sector resulting from conversion of land from agricultural to urban use, a period of low farm prices, and some retirement of agricultural land from any active use. On an absolute basis, no progress has been made in reducing industrial groundwater use.

Table13.2 Arizona Groundwater Use by AMA and Sector

	Phoenix AMA		Prescott AMA		Tucson AMA	
Year	1985	1998	1985	1998	1985	1998
All Sectors						
Total AF Km^3	1,020,545 12.58	677,427 8.35	17,388 0.21	15,369 0.19	277,447 3.42	298,700 3.68
% GW	49	34	89	82	78	87
Municipal						
AF Km^3	276,541 3.41	196,112 2.42	5,323 0.07	10,304 0.13	111,055 1.37	150,800 1.86
% GW	42	23	100	99	97	94
Agriculture						
AF Km^3	681,438 8.40	402,378 4.96	11,988 0.15	4,489 0.06	111,004 1.37	68,853 0.85
% GW	50	40	85	63	96	74
Industrial						
AF Km^3	62,558 0.77	78,937 0.97	77 0.001	576 0.007	55,388 0.68	56,100 0.69
% GW	75	48	100	100	99	99

Source: Governor's Water Commission, Technical Advisory Committee, Draft 10/16/2000.
Table omits water withdrawn from exempt wells or on Native American reservations.

The GMA required each of the AMAs to develop detailed water management plans that were to be revised every five years. When the first management plans were developed, the primary requirement imposed on municipal providers was the gallons per capita per day (gpcd) program. The premise of this was that municipal use could be reduced by requiring each provider to work toward a pre-determined goal. However, both the setting of individual goals and the measurement of progress have been subject to wide disagreements. The service area of each provider differs by the historic use patterns resulting from factors such as lawn sizes, the availability of non-municipal irrigation water or private wells, subdivision vegetation regulations, proportions of residential users in single family houses and apartments, and the presence of water intensive commercial and industrial activities. Despite lots of publicity and other programs such as free low flow showerheads, little progress has been made to date in reducing municipal gpcd.

13.6.1 PHOENIX AMA

In spite of rapid urbanization in the Phoenix AMA, agriculture still uses more total water than any other sector. The AMA has made progress in substituting CAP water for groundwater. The first two management plans emphasized increasing supplies of water and augmentation programs. Most of these opportunities have now been exhausted and the Third Management Plan places increased emphasis on recycling of water and demand management through regulation.

The Phoenix AMA's most important contribution to the water management process may be its extensive research on indoor and outdoor residential water use. Through study of residential users, the AMA has found that larger volume residential water users have more potential for conserving water than small water users. For example, residential users with indoor use of less than 57 gpcd (216 liters) and outdoor use of less than 117 gallons or 443 liters/household/day were determined to have no potential for reduction in their water use while those using more than 87 gpcd (329 liters) and 256 gphud (969 liters) respectively were determined to have maximum potential (Phoenix AMA, 1999). This knowledge is useful in designing pricing systems for water and also may bear on equity of water use pricing and /or restrictions as higher income residential units tend to have the higher water uses. It also signals the potential use of demand management as an effective tool in limiting water use.

Augmentation is heavily emphasized by the Phoenix AMA plan, since this area already imports both groundwater and surface water from sources outside the AMA. While importation of groundwater is generally prohibited by the GMA, Phoenix is well positioned to take advantage of several exceptions to this policy (Phoenix AMA, 1999). Locally produced treated effluent amounting to353 million or 286,000 acre feet (af) per year is available in the Phoenix AMA. An acre foot is 325,851 gallons, 1,233 m^3 or one acre covered by one foot of water. Currently, the largest customer for reclaimed effluent is the Palo Verde Nuclear Generation Station with an annual use of 56,000 AF (69 million m^3). The Phoenix AMA is gradually finding more ways to utilize treated effluent, as well as building recharge and storage facilities for this water.

Achieving the goal of safe yield by 2025 will be difficult at best. Adjudication of Native American water rights, decisions by the CAP managers, the rate of decline in agricultural irrigated acreage, and the associated transfer of water rights to newly urbanizing areas are factors over which the Phoenix AMA has little control. While municipal use is subject to a renewable supply requirement, Native American and industrial water use is not. Hydrologic and economic models have been developed which provide extensive knowledge of how safe yield might be reached. The challenge is to turn this scientific knowledge into effective public policy.

13.6.2 PRESCOTT AMA

It was not until January 1999 that the Prescott AMA was determined to be not at safe yield. The AMA lacked adequate hydrological data to make this determination at an earlier date (Prescott AMA, 1999). The Third Management Plan for the Prescott AMA focuses on strengthening the mandatory conservation requirements and augmentation programs such as the use of treated effluent, purchases of surface water rights from the Chino Valley Irrigation District, and importation of purchased groundwater from retired farmland in the Big Chino Basin (Prescott AMA, 1999).

Prescott is too far from the CAP canal to economically use this source of water. Prescott has already transferred the rights to its share of CAP water and used the proceeds to develop other sources of water. While Prescott is planning to use some water from the Verde River to supplement its ground water supplies, negotiation is necessary with Phoenix and other cities to the south for rights to this water.

Agricultural groundwater use is declining in the Prescott AMA but municipal and industrial use is increasing. Thus, a focus on conservation requirements is important to the ground water use reduction program. Dry lot subdivisions, where homeowners use exempt wells (less than 10 AF/yr.), are proliferating. However, in the Prescott area, many of these wells are expected to go dry as groundwater levels continue to fall. This will probably result in an extension of municipal water service to these areas, subjecting them to the conservation rules of the Prescott AMA.

13.6.3 TUCSON AMA

The Tucson AMA's goal of safe yield is not expected to be met in 2025 (Tucson AMA, 1999). This is the result of several problems in the Ground Water Management Act itself and associated uncertainties. While achieving the management plan's goal depends heavily upon reducing irrigation on land with grandfathered groundwater pumping rights, under current law the AMA cannot regulate irrigation water use on 15,600 acres on the Tohono O'dham Reservation (Tucson AMA, 1999). Under current plans, the tribe will obtain the bulk of its water under contract from the CAP while an additional 28,000 acre feet granted to the tribe from the Arizona Water Rights Settlement Act of 1982 will be in the form of exchange water of appropriate quality (Tucson AMA 1999). A second problem is that the grandfathered irrigation water rights exceed the total agricultural water use in the plan. New legislation reducing the amount of water associated with these rights will be necessary to achieve the goal of the

Management Plan (Tucson AMA, 1999). This will be difficult or impossible to achieve since Arizona's legislature is very protective of existing property rights.

Mining has rights to 62,000 acre feet of water in the AMA but is only expected to use about 45,000 acre feet by 2025. An upturn in the demand for copper or discovery of a more competitive ore body could cause mining to increase its pumping of ground water resulting in missing the stated goal by an even a higher amount.

In the early 1990s, there were one hundred and twenty-six small water providers (those serving less than 500 people or less than 100 AF or 123,275 m^3 per year) in the Tucson AMA. While these small producers are exempted from regulation, it is important to consider that if each of these providers pumped the maximum allowable amount of water, an additional 12,600 AF (15.5 million m^3) of groundwater could be pumped that is not subject to the GMA. Purchase of these smaller providers by larger nonexempt water providers or mergers forming larger nonexempt providers could reduce this problem. The Third Management Plan (Tucson AMA, 1999) is the first to consider the use of treated effluent as a water source for irrigation, mining, or industrial use. However, a number of farms downstream of the sewage treatment plant in Tucson continue to benefit from effluent dumped into the Santa Cruz River. Further use of effluent could make a substantial contribution to meeting the goals of the AMA.

It is legal in Tucson to dig wells for personal use on private land within the water provider's territory. A number of home-owners recently dug or are digging wells in the upscale El Encanto Neighborhood to provide water for the lush landscaping that is typical of this subdivision. They hope to avoid increasing water rates charged by Tucson Water and assure themselves of a supply that is unregulated. Since these wells are not controlled by the local AMA, the city or the water provider, the amount of water pumped by them serves to offset conservation gains made by the activities of these entities and the other citizens of the city. For effective conservation planing, individual pumping should be strictly regulated or be made illegal in areas within an AMA that are served by a recognized water provider.

13.7 Exempt Wells

In Arizona, certain wells are exempt from regulation. Exempt wells are those that pump less than 35 gallons (132 liters) a minute and less than 10 AF (12,327 m^3) per year for domestic purposes, stock watering, and/or irrigation of less than two acres of land. Table 13.3 shows that the estimated quantity of groundwater pumped by these wells in the Tucson and Prescott AMAs has more than doubled between 1985 and 1998. These unregulated groundwater withdrawals interfere with the ability of the Tucson and Prescott AMAs to meet their safe-yield objectives.

Table 13.3 Exempt Well Groundwater Use in Urban Arizona AMAs

AMA \ Year	1985	1998
Phoenix AMA	12,285 AF (151 million m^3)	11,000 AF (136 million m^3)
Prescott AMA	732 AF (9 million m^3)	1,501 AF (19 million m^3)
Tucson AMA	1,000 AF (12 million m^3)	2,250 AF (28 million m^3)

13.8 AMAs Replacement Water Options

One theme sounded throughout the Groundwater Management Act and the individual management plans is the replacement of groundwater with surface water, treated effluent, or water from distant sources. The Tucson AMA does not have readily available sources of nearby surface water and is finally making use of CAP water. While Phoenix may have some additional surface water available from the Salt and Gila Rivers, this region also uses CAP water to sustain its economic and population growth. Thus, much of the counted on replacement water is sourced from the Colorado River via the CAP. Both surface storage in reservoirs and underground storage in aquifers are being used to accommodate differences in timing of availability of Colorado River water and urban demands. Storage is also designed to provide continuity of supply when river water is not available due to repairs of the delivery system or other problems. Future development on the Colorado River, effective efforts by Mexico to obtain more water for the Colorado River delta area which has been harmed by reduced flows, and the settlement of Native American Water Rights claims by provision of Colorado River water may reduce the future availability of CAP water.

An alternative source of replacement water is groundwater or surface water that lies in Arizona but outside the AMAs. Water purveyors within the AMAs have purchased rights to Arizona water outside of the AMAs, usually from agriculture. Future economic development of rural areas may be restricted by the loss of water rights (Briggs, 2000). Arizona is still developing economically and a number of water source areas outside the AMAs have great potential that is likely to be important to the state economy in the future. There is a danger that conveying this rural water and the associated water rights to the urban AMAs may provide fewer benefits for the future Arizona economy than using it in its original location.

Effective coordinated management of all water sources – groundwater, surface flows, imported water and reclaimed water – is severely impeded by lack of a single unified water management agency and differences in the laws governing each type of water. Since surface and ground water are hydrologically interconnected, the statutes should be revised to allow planners and managers to better coordinate the use of these water sources.

13.9 Conservation Policy

Mining, agricultural, and industrial demands for water are managed within the AMAs by direct regulation and negotiation between the AMA and the individual user. Thus, adjustments can be made in the allowable pumping by the direct users. In municipal use, the water provider (e.g.: Tucson Water) is responsible for the required reductions in use. Each municipal supplier, in consultation with its AMA, must devise a strategy to reduce water use in the face of a variety of legal, economic, and political constraints.

13.9.1 WHO IS RESPONSIBLE?

A number of policy makers and analysis have recently advocated shifting some of the responsibility for water use reductions to the individual municipal water customers and away from the municipal providers. The thinking is that the final user is better able to control use than the provider. This would be a break with the current policy of making the municipal water right holder responsible for water conservation efforts. Requirements that might be imposed include direct limits on total water use per resident or employee, use of low flow devices, leak checks, limits on overflows, restrictions on the kinds and number of plants used in landscaping, and swimming pool covers to reduce evaporation. Each of these requirements intrude on individual privacy and rights to choose as well as being expensive to monitor and enforce. Since water companies do not have police power, such limits would necessarily be imposed and enforced by some governmental agency.

13.9.2 MARKETABLE PRIVATE WATER RIGHTS

An alternative to direct government regulation of end uses of water is privatization of water rights. To result in safe yield, the Arizona Department of Water Resources (DWR) would determine the annual amount of water that could be pumped in each area within each AMA and then distribute a corresponding number of water rights. These would most likely be adjusted forms of existing water rights. The allowable amount of pumping for each right could be varied from year to year as needed to achieve safe yield. A market could be established for these rights so that holders could buy and sell them in accordance with their water use demands. Illegal water withdrawals would be subject to penalty rates substantially higher than current market prices for water. A major challenge in establishing a system of marketable water rights is the initial distribution of the rights. If individual residential customers are included in the program, there would be a huge number of rights holders. Would rights be distributed to current water users only? What about owners of vacant land? If vacant land were excluded, its market value would fall dramatically, and new housing and commercial development would be difficult and more expensive, since water rights would need to be purchased before the land could be used. Being required to assemble the water rights for a housing development or corporate expansion might encourage the entrepreneur to invest elsewhere. Monitoring costs, contract enforcement, and transactions costs are likely to be very high if the market includes individual residential customers. These drawbacks would make a trading scheme unappealing to politicians who would have to support the water market.

13.9.3 RAISING WATER PRICES TO LIMIT DEMAND

Only a small fraction of urban water is used for drinking, washing, and other necessity uses. Therefore, higher water prices can be effective in reducing demand. In contrast to regulations, higher prices do not intrude on rights and privacy or imply monitoring and enforcement costs. Given a water rate schedule, each customer decides how much water to use. Furthermore, a penalty rate could be imposed for very high (excessive) use within each customer category. However, large price increases are politically unpopular and may be viewed as unfair by a large number of customers. A large price increase in Tucson in 1976 led to the recall of the city council, and a general unwillingness to consider using price to regulate demand (Martin, et al., 1985).

Proper pricing requires periodic rate increases to adjust for inflation and changing local income levels. A simple way to achieve this result is to index water rates to the Consumer Price Index (CPI). The implementation of an annual automatic CPI adjustment would likely require considerable time, political risk, and effort on the part of those responsible, but once in place would no longer be an issue.

Water price has not been a major weapon in the AMAs as a strategy for reduction of overdraft, nor has it been used effectively by individual municipal providers. A typical residential customer, consuming 7,500 gallons (28.4 m^3) per month, saw very little increase in their inflation-adjusted bill for the period from 1993 to 1997, as shown in Table 13.4. In some AMA cities, inflation adjusted prices fell during this period. Only Gilbert and Mesa, both in the Phoenix AMA, increased their inflation adjusted water bill for 7500 gallons by more than $2.00 over the time period.

Fuller (1998) found that the price elasticity of demand for water is lower in Phoenix (-0.36) than in Tucson (-0.80). Fuller attributed this difference to the higher income per family and lower price of water in Phoenix. The Water Infrastructure Finance Authority of Arizona (1999) found that 7,750 gallons had a commodity cost of $11.93 from Tucson Water and only $5.61 from the City of Phoenix. During 1995, Phoenix municipal customers used 230 gallons (871 liters) per day (Phoenix AMA, 1999) compared with 167 gallons (632 liters) per day in Tucson (Tucson AMA, 1999). In the early 1970s, before Tucson prices increased, usage in Tucson and Phoenix was similar.

13.9.4 INCREASING BLOCK RATES AND SUMMER SURCHARGES

The majority of communities in the AMAs have introduced increasing block rates and the Arizona Corporation Commission has approved them for use by private water companies as long as they do not result in increased profit. Increasing block rates serve residential customers well as a means of providing equity in water rates between income levels and between user levels. The higher volume residential users of water tend to be high-income families who can afford water-using devices such as multiple bathrooms, swimming pools, lush lawns, decorative fountains, and mist systems. A mist system puts out a fine spray of water in an outdoor patio area to provide evaporative cooling. Lower income people do not have these extra demands for water and, therefore, pay a lower marginal price for water.

Arguing that Tucson's residential block rate structure did not result in great water savings, Martin and Kulakowski (1991) point out that poor block rate design may make these complex rates ineffective. A structure with only a few blocks and large jumps in price between blocks would have the greatest impact. They suggest that summer surcharges probably contributed more to the reduction in consumption experienced in Tucson than the block rate structure. Black and Veatch (1997) report that Phoenix, Chandler and Glendale in the Phoenix AMA, along with Tucson, utilized some type of summer surcharge. Because of extensive residential lawn watering, these utilities experience a large summer peak usage which has been reduced by higher summer water rates.

Table 13.4 Typical Arizona Residential Water Bills
7500 Gallons (28.4 m^3) per Month for Selected AMA Cities.

Water Provider	Price in Current Dollars			Inflation Adjusted Price		
	1993	1995	1997	1993	1995	1997
Phoenix AMA						
Avondale	$15.58	$17.00	$17.00	$10.92	$11.31	$10.69
Chandler	$16.19	$18.06	$18.06	$11.35	$12.02	$11.35
Gilbert	$11.98	$16.05	$17.08	$ 8.40	$10.68	$10.73
Glendale	$16.22	$17.21	$17.74	$11.37	$11.45	$11.15
Mesa	$ 7.50	$15.39	$16.44	$ 5.26	$10.24	$10.33
Peoria	$11.95	$14.40	$15.36	$ 8.38	$ 9.58	$ 9.65
Phoenix –winter	$ 8.75	$ 9.79	$10.00	$ 6.14	$ 6.51	$ 6.29
Phoenix-summer	$ 5.43	$ 6.23	$ 5.96	$ 3.81	$ 4.14	$ 3.75
Scottsdale	$18.54	$20.92	$21.11	$13.00	$13.92	$13.27
Tempe	$13.45	$14.50	$14.19	$ 9.43	$ 9.65	$ 8.92
Prescott AMA						
Prescott	$20.30	$20.30	$20.30	$14.24	$13.51	$12.76
Tucson AMA						
Tucson-winter	$15.85	$15.99	$16.54	$11.12	$10.64	$10.40
Tucson-summer	$16.35	$15.99	$16.54	$11.47	$10.64	$10.40

Deflated by the United States Consumer Price Index (1983 = 100), U.S. Department of Commerce, Bureau of Labor Statistics. Source: Black and Veatch (1993, 1995, 1997).

13.9.5 COMMODITY TAXATION TO LIMIT DEMAND

Raising the price of water is simpler and often more effective as a method for reducing water use than direct regulation. Providers, however, are ill prepared for this alternative because of political opposition to higher water rates, Arizona Corporation Commission (ACC) reluctance to allow regulated providers to raise rates and a general belief that price is not important. The State of Arizona could eliminate these barriers by imposing a commodity tax on each gallon of water withdrawn from the aquifer. Political opposition could be attenuated by using the revenue to directly reduce county property taxes. Alternatively, the tax revenue could be used to subsidize CAP and other renewable water supplies. The tax rate could vary with the observed overdraft in each sub-aquifer within each AMA. Where water tables are falling rapidly, the tax would be quite high. Where they are stable or rising, rates could be low or even zero. Such a tax would eliminate the conflict experienced by private water providers in the AMAs between regulations imposed by the AMA and the water rates allowed by the ACC. The tax would also give an incentive to move water from where its value (tax rate) is low to where its value is high. For example, an effective commodity tax might encourage shipment of water from southwest Phoenix where water tables are rising to northern metro areas where excess withdrawals are occurring.

An effective commodity tax on mined groundwater could also help solve the problem of incorrect relative prices for water of different qualities. In many areas of Arizona, high quality groundwater can be used for the often very low cost of pumping while lower quality CAP water is generally more expensive and treated effluent, which is of the lowest quality, costs even more. A commodity tax on pumpage of high quality groundwater could be used to subsidize the use of low quality treated effluent.

13.10 Commodity Value of Water in Arizona

In general, legal restrictions and public policies in the state do not allow for the commodity value of water to be considered in setting water rates. As long as raw water is regarded as a free good there is little incentive to conserve it and the externalities caused by over drafting are likely to continue. No payment is made for groundwater itself. In most instances, groundwater users pay only for pumping and delivering this water. Of course, if the water provider must go to distant sources to get water and purchase water rights, some market value of the water will be included in water charges. The Tucson and Phoenix plans to replace groundwater with CAP water from the Colorado River include paying more for water from this source. However, the price of CAP water does not come close to the actual cost of transporting the water thanks to a federal subsidy for a substantial share of the water provision infrastructure. The commodity value of Colorado River water is generally ignored. Most of the rights to Colorado River water are determined by compact among the states located along the river and not by free market bidding which would establish a market price for water. Furthermore, the total amount of water allocated by this compact exceeds Colorado River flows in draught periods. However, the price charged for water has not been raised in these periods to ration the reduced availability of Colorado River water.

13.11 Concluding Remarks

Arizona, with its rapid population growth, will continue to have water problems that cannot be easily fixed. Groundwater is being mined and exhausted and there is a limited amount of CAP water. Full utilization of effluent for non-potable uses could facilitate continued expansion for many years. However, when CAP allotments and effluent are fully utilized, economic and population growth could be limited unless newer and more expensive sources of water such as desalinized seawater are developed. While the AMAs are in place to help solve future water problem when they occur, the political will on the part of the legislature to effectively deal with these problems has been sadly lacking. Specifically, laws must be enacted to deal with the problems caused by exempt wells, incompatibility of current groundwater and surface water regulations, and requirements for substitution of recycled water for potable water when possible.

References

Black and Veatch, 1995, 1997, 1999. *Arizona Water/Wastewater Survey,* Phoenix, Arizona.

Briggs, P., 2000. "Can Water Supplies Keep Up with Population Growth?" unpublished paper presented at 20th Anniversary Symposium of Arizona's Groundwater Act, Tempe, Arizona, May 1-2.

Carpenter, M.C., 1999. "South Central Arizona: Earth Fissures and Subsidence Complicate Development of Desert Water Resources," in D. Galloway, D.R. James, and S.E. Ingebritsen, eds., *Land Subsidence in the United States,* U.S. Geological Survey, Circular 1182: 65-81.

Fuller, J.R., 1998. *Financing the Central Arizona Project,* Ph.D. dissertation, University of Arizona.

Gelt, J., 1992. "Land Subsidence, Earth Fissures Change Arizona's Landscape," *Arroyo,* 6(2): 1-4.

Martin, W.E., and S. Kulakowski, 1991. Water Prices as a Policy Variable in Managing Urban Water," *Water Resources Research,* 27(2, Feb.): 156-166.

Martin, W.E., H.M. Ingram, N.H. Laney, and A.W. Griffin, 1985. *Saving water in a Desert City.* Resources for the Future, Baltimore.

Phoenix AMA, 1999. *Third Management Plan for Phoenix Active Management Area.*

Prescott AMA, 1999. *Third Management Plan for Prescott Active Management Area.*

Tucson AMA, 1999. *Third Management Plan for Tucson Active Management Area.*

Water Infrastructure Finance Authority of Arizona, 1999. *Arizona Water and Wastewater Residential Rates-1999 Survey,* Phoenix.

Chapter 14

Urban Water in Israel

Yoav Kislev
Hebrew University of Jerusalem

14.1 Introduction

Israel is a small country in a semiarid region. Its economy is industrialized and modern, its income is comparatively high. The water economy of Israel is also modern. It is founded on a law that specifies central control of the resources, and in general, water provision and sewage removal are adequate. Urban consumers receive an ample supply of water of reasonable quality. There are, however, more than a few weak points in the structure of the urban water sector and the water economy in the country at large; they will be highlighted together with reforms that are being introduced in an effort to amend shortcomings.

Israel shares water sources and watersheds with its neighbors, particularly with the Palestinians on the West Bank and the Kingdom of Jordan on the east. This survey deals only with Israel proper, although several Israeli urban settlements on the West Bank are included in the official municipal statistics quoted in the chapter.

14.2 The People and the Economy

The State of Israel was established in 1948 following a United Nation resolution on the partition of Palestine. The war between the emerging nation and the countries around it

D.E. Agthe et al. (eds.), Managing Urban Water Supply, 219–240.

Figure 14.1 Map of Israel and the National Project
Source: Kliot, Nurit, 1994

ended with Israel occupying a larger area than was allotted to it in the resolution, but its borders were not finalized and have since been referred to as the "green lines," the color used on the ceasefire maps. The green lines and international borders are marked in Figure 14.1—depicting the northern two-thirds of the country—by segmented lines separated by dots. In another war, in 1967, Israel occupied all of the area of Palestine west of the River Jordan. More recently, some of the occupied areas were handed over to the Palestinian Authority in accordance with various stages of peace agreements signed since 1993.

It was estimated that 650,000 Arab refugees left the war regions in 1948 and moved to neighboring areas. The population of Israel, which numbered less than a million in 1948, doubled in size in its first decade, mostly due to large Jewish immigration from Europe and Middle Eastern countries—many of them refugees. Another large wave of immigration, again close to a million people, came in the 1990s in the wake of the collapse of the Soviet Union. The population's average growth rate for the last half-century was 4% per year and by 2002 the total number of people residing in the country has reached 6.5 million.

More than 90% of the people of Israel live in urban areas. Agriculture employs less that 2.5% of the labor force and this share has been declining for a long while. Despite the decline, rural communities have been expanding as many who exited agricultural occupations stayed in their rural homes which, in a small country such as Israel, were never too far from urban centers of employment and commerce. As the population grew, municipalities were upgraded. A large number of today's towns and cities were regarded as villages 40 and 50 years ago.

In its early period, Israel was recognized as a developing country, with a comparatively low income level. Since then, per capita Gross Domestic Product increased on average 3.3% per year, partly thanks to international aid. At present GDP is approximately $16,000 U.S. per person. The economy is modern and the standard of living is close to European.

14.3 The Water Sector

Israel is a small and narrow country; half of its area is desert. Precipitation, only in the winter, averages more than 700mm per year in the north and less than 35mm in the southern tip of the country. Throughout the country's history, its people relied on springs, rivers, and shallow wells for water provision. Intensive utilization of water resources started in the second half of the 19th century with the expansion of the citrus industry in the coastal plain. Up to the 1950s, water projects were local in nature. An integrated national system was established in the early 1960s with the completion of the National Water Carrier connecting the Sea of Galilee (Lake Tiberias) in the north with water sources and points of demand in the center and the south (the carrier is marked as Kinneret-Negev Conduit in Figure 14.1).

The largest utility, Mekorot Water Co., supplies two thirds of the water in Israel and also operates the National Carrier. Mekorot is a government company; while other suppliers are private well owners, municipalities, and regional cooperatives.

Municipalities are required to collect and treat their sewage and several cities have cooperative projects with agricultural interests in their vicinity.

The "safe yield" water supply from Natural sources is estimated as 1,550 million m^3 per year (Water Commission, 2000). The first comparatively large, 100 million m^3 per year, desalination plant is soon to be constructed on the coast of the Mediterranean south of Tel Aviv and preparation are under way for the installation of additional capacity and for the import of 50-100 million m^3 a year from Turkey in converted oil tanks.[34] Currently, 270 million m^3 of recycled water are utilized in Israel and the forecast is that by the year 2020 the country will make use of 830 million m^3 per year of recycled water. Consumption of fresh water is some 720 million m^3 in households and industry and the rest in agriculture. The government aims to supply agriculture with 1,160 million m^3 per year of water from all sources (including reclaimed and brackish water) but this may not be possible in the coming years.

In Israel's early days, in the 1950s, total water supply amounted to less than 600 million m^3 per year. Capacity was increased to irrigate dry lands and to expand and improve supply to the growing population. However, resources were limited and their development expensive; population growth has surpassed water supply over the last half-century, and the amount available per person (in the country's population) has declined. By the data depicted in Figure 14.2, per-capita consumption in industry has remained essentially constant for the last 40 years, while consumption in households increased at an average yearly rate of 0.66% (calculated in a regression of log consumption on time). Over the same period, per-person water available for agriculture was cut in half compared to the volume of the 1960s. Despite the reduction, agricultural production per-capita is, today, more than 150% of the quantity produced 40 years ago. Therefore, water productivity in agriculture has increase threefold over the period. A similar increase was also recorded in industry. Water potential and safe yields are generally stated in terms of average use. Precipitation varies from year to year and, with fluctuating rains, natural replenishment of the reservoirs is not stable. There is a tradeoff between average supply and its reliability. With a policy of regular extraction of large quantities of water, reservoirs are often low and a reliable supply cannot be maintained. This truism was brought home twice in the last 15 years, once in 1990-91 and again more recently. Israel is facing now, as it also faced more than 10 years ago, an acute water crisis into which the country slid when several dry years followed a period of over-utilization. The crisis caused a public outcry and even panic. And indeed, the water level in the Sea of Galilee is lower now than it has ever been in known history and the aquifers show clear signs of overdrafting. The recent crisis and the panic it caused hastened Israel's move to desalination and importation.

[34] Israel has always been importing large quantities of virtual water (a term due to Allen, 2000) in grain and beef purchased in water rich countries.

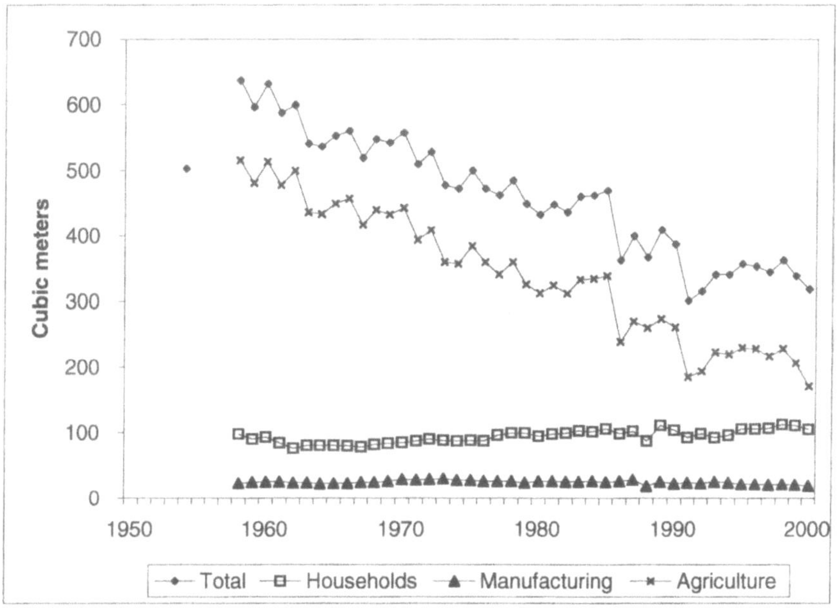

Figure 14.2 Per-capita Water Utilization in Israel
(Water use divided by the population in the country)
Source: Central Bureau of Statistics, *Statistical Abstract of Israel.*

14.3.1 PUBLIC CONTROL

As natural resources, water reservoirs are common pools. Under open access, individuals will behave as free riders: they will extract water as long as it is beneficial for their own use disregarding the detrimental effect that their action may have on other users of the reservoirs (for example, by lowering water levels or drawing in salty ocean water). The common resource will be depleted (Hardin, 1968). In addition, suppliers are monopolies, particularly Mekorot. These features call for government intervention. Consequently, "The sources of water in the State are the property of the public, are given over to the control of the State ..." (Israel Water Law—1959) there is no private ownership of water in Israel. The Water Commissioner is responsible for the utilization and the sustainability of the resources. The law requires metering of all uses of water, which means that wells and pumps are monitored and consumers—households, manufacturers, farmers, and others—pay by the quantity they use.

14.3.2 ALLOCATION

There are two major water allocation problems:

- allocation of extraction; where, when, and how much to pump
- allocation of water for utilization and consumption

The two problems are distinct, although the wording of the law obscures the distinction. The criterion for extraction of water is sustainability of the resources. The role of the Water Commissioner is to guard the long run stability of the quantity and quality of water. Fulfilling this role may require decisions on each source and well separately, depending on local hydrological circumstances. The criterion for the allocation of water for consumption and utilization is efficiency; that is, maximization of economic welfare from the use of water. Two management instruments are in use: prices and quotas. Households and most manufacturers can purchase all they demand at the established prices from Mekorot. Water in agriculture is allocated by quota and, in addition, farmers who purchase water[35] pay Mekorot or the regional suppliers.

As water is the property of the public and controlled by the state, suppliers or private well owners are issued extraction licenses together with allocation instructions showing how much to provide each user. In principle, both the licenses and the quotas are reallocated every year, but in practice they seldom change. Possession creates rights and as a result the water sector operates under a mixed, sometimes fuzzy, regime of public ownership and private rights. For example, although the law does not explicitly prohibit trading in water, it does not permit it either. Officially the Water Commissioner was against trading in quotas, but in many cases he allowed the exchange. However, an official water market has not been established yet.

The multitude of possibilities was also reflected in the courts. A milestone decision was handed down in 1964 when a farming community in the coastal plain refused to contribute its high quality water to the National Carrier in exchange for more salty Sea of Galilee water. The Supreme Court ruled that, by the law, the water commissioner may force a water allocation program, including the mixing of water of different qualities. In another case, in 2001, a court did not accept the doctrine of the omnipotent commissioner and rejected his decision that, since the water—all of it—is public property, cities cannot sell recycled water to farmers and the water they treated must be handed over for irrigation free of charge. (Indeed, in practice most cities do not charge for water leaving their recycling plants.)

14.3.3 HYDROPOLITICS

Government intervention in the management of water is a necessary consequence of the common pool features of the resources and the monopoly position of the suppliers. Once the government is involved in any issue, interest groups rise in an attempt to

[35] Some farmers own private wells and others, around and above the Sea of Galilee, pump directly from the lake or the Jordan River (under license).

change public policies in their favor (Zusman, 1997). The two strongest groups in the water sector, the farmers and the workers of Mekorot, have different interests and attempt to affect different aspects of policy. The "greens" form the third group. They are growing in strength but their overall effect is still marginal (the green assessment of the water sector of Israel is summarized in Auerbach and Karassin, 2002).

The farmers push for larger water provision at the lowest possible price. Indeed, the development of agriculture was uppermost on the minds of the builders of the National Carrier and its extensions. It seems, however, that in more recent years lower world food prices, the declining share of agriculture in the national economy, and the increased demand for water in the urban sector reduced the power of the agricultural lobby. Compared to farmers in other countries, farmers in Israel pay high prices for irrigation water. Mekorot charges 0.20-0.25 U.S. dollars per m^3, while independent suppliers and regional cooperatives pay an extraction levy which will be explained later. Consequently, in recent years, prices, not quotas, determined water utilization in agriculture. Still the farm lobby was strong enough to guard the policy of overdrafting that resulted in the acute crises of the last 15 years.

Mekorot workers are interested, naturally, in their positions and income. They protect their interests by defending the monopolistic position of the utility. The government has been attempting for years to reform the structure of the company and to reduce its market and political power. Any achievement in this endeavor was despite a strong opposition of the workers and their union. We shall encounter traces of this struggle in the discussion of the urban water sector below. The ongoing conflict between Mekorot and the government, its official owner, uncovers the real governance structure of the company—it is run by its workers much more than by its owners. It should however be added that the workers are a dedicated lot. Water is not cheap but Mekorot's provision is reliable and of good quality.

Contrary to the farmers and Mekorot's workers, the public at large is not organized and its voice is not heard when decisions on the utilization of the resources are made. But it is not completely overlooked. For example, the recent decision to move to desalination despite the heavy investment the new technology entails is to a large extent a reaction to public pressure that the government "do something" in the face of the water shortage. And, despite the shortages and curtailed provision to agriculture in the recent crises, there were no cuts in supply to urban consumers. The only way urban dwellers felt the shortage, besides being exposed to alarming stories in media, was through the promotion of a water saving campaign. The government did not approve the suggestion of the Water Commissioner to prohibit irrigation of parks and to raise marginal price of water to heavy residential consumers. It probably figured that it would rather risk the health of the aquifers than face discontent among noisy urban voters.

The government, although taken generally as a unified entity, does not always speak in a single voice. In particular, the Treasury is concerned with conservation of the fisc and with economic efficiency. Other offices are often striving to promote the interests of the sectors they are responsible for (sometimes seem to represent) and projects dear to their hearts.

14.4 Early Urban Supply

The first human settlers kept close to sources of water. Natural supply was supplemented by rainwater artificially collected in cisterns dug in the ground and sealed to reduce leakage. In towns, water was diverted to the cisterns from rooftops and courtyards; in the desert, occasional floods were stored in nearby cistern and sometimes carried later to storage facilities close to home. Biblical Jerusalem, King David's city, was built on a slope reaching down to a spring in the valley underneath. In the 8th century B.C., to prepare for emergencies, King Hezekiah "constructed the pool and the conduit to bring water into the city" (2 Kings, 20, 20). The spring and the tunnel are still drawing admiring tourists. Water collected from springs, wells and cisterns was augmented by water carried from distances. Ancient Roman aqueducts, built some 2000 years ago, can still be seen in several places in Israel. A few of the aqueducts were maintained and rebuilt but many were neglected and fell into ruin.

Gravitational water conveyance and distribution systems were the hi-tech projects of the ancient world and technology did not change much for generations. As late as in the early 19th century, when the need arose to expand supply of water for the city of Akko (Acre, north of Haifa), the solution was a new aqueduct. This one may, however, have been the last.

Where groundwater was available, wells were dug by hand and the water was lifted by human or animal power. These wells could not be deep and they did not support large flows. Engines and lift pumps replaced animals at the end of the 19th century and the beginning of the 20th. The first reports of mechanical well drilling in Israel are from the 1920s (Karlinsky, 2000).

14.5 The Municipal Sector

There are 263 municipalities in Israel classified by their governing bodies: 64 cities with city councils, 146 towns with local councils and 53 regional councils. Cities are generally larger than towns.[36] Regional councils are comprised of villages and other small places of dwelling and do not include communities of urban character. Water supply to residential users in the regional councils is, in many cases, part and parcel of the water supplied to agriculture and is not measured separately. I shall therefore deal in the survey only with cities and towns.

The municipalities are responsible for local services; including education, welfare, roads, parks, water, sewerage, and garbage collection. In education and welfare they act as agents of the national government that covers most of the cost of services specified in the law of the land. In other lines of activity the municipalities are expected to operate independently and to finance their services from local taxes, mostly on property, and specific charges (such as for water). However, the national government is nevertheless deeply involved in local affairs—on average, a third of the outlays of the municipal governments comes from the national budget. The involvement is motivated by the

[36] The terms city and town are the author's, the official terminology refers to the governing body of the locality.

desire to assist weaker communities and to enhance the development of the country, and also, evidently, by the conviction that big brother knows best. Also, assistance to local authorities, delivered at the grass roots, is deemed to have high political payoffs; parties in coalition governments vie for control of the Ministry of the Interior, the supervisor and supporter of the local authorities. Whatever the motivation behind it, government interference in local affairs creates a situation of "soft budget constraints"—mayors can spend beyond their means and plead with the government to cover deficits.

The municipalities differ significantly by income of their residents. The 1999 average for the 10 poorest communities (out of 186 cities and towns for which data were available) was less than 3000 U.S. dollars per person per year. The richest ten registered more than $14,000.[37] As usual, the poorest communities have the largest number of children. In most, the share of children of 0-17 years old exceeds 45% of the population while in rich communities the corresponding shares are 30% and lower. The meaning of these numbers is that the weaker municipalities are torn between the need to provide vital services, particularly schooling, on a comparatively large scale and the poorer ability of their populations to support them. Naturally, however, municipalities in higher income communities are also short of funds for project they would like to promote, a fact that often has (particularly had in the past) a significant bearing on their water economies.

Special problems are posed by minority municipalities. Of the population of Israel, close to 5 million are Jews and the rest, 1.5 million, are minorities, mostly Arabs. The history of war and the continuing tension between Israel and its neighbors has naturally reflected on the relations of the state with its Arab citizens. Seven cities in Israel have both Jewish and Arab communities—among them the largest three: Jerusalem, Tel Aviv, and Haifa—but most of the localities are separated, either Jewish or Arab (77 municipalities are Arab). As a rule, the Arab communities are poorer than most Jewish municipalities and they also have less developed water and, particularly, sewage and recycling systems. This gap is due to neglect on part of the government, to the fact that Arab towns and cities are traditional villages that grew by leaps and bounds without adequate town plans, and to lags in the collection of charges for water and sewage removal in Arab communities—they lack therefore the ability to participate in government supported program based on matching of local and national funds.

However, governments in democracy cannot keep neglecting more than a fifth of the population and its voters. The water and sewage sector of the Arab municipalities has been improving, even if slowly, and can be expected to reach the standards of the country at large in the future.

14.6 Urban Water

Several municipalities rely to a large extent on their own water sources; some even supply more than 50% of their needs. Most others have to get all their water from the national utility. For the country as a whole, Mekorot supplies more than 80% of the

[37] The difference is exaggerated somewhat as welfare payments are included in income but income taxes are not deducted.

urban water. In a few exceptional cases urban communities buy water from regional agricultural cooperatives.

Towns and cities received and extracted 614 million m^3 of water in the year 2000. Table 14.1 reports the distribution of uses by sector. Most municipal water goes to urban uses but small amounts are provided to agriculture and industry located within municipal areas. Urban water is divided into sub-sectors, with the largest quantities consumed in households. Overall, the residential sub-sector's share of municipal water is 56.7%.

The average residential water consumption in the municipalities in the year 2000 was 60.5 m^3 per person while total urban water consumption was 86.1m^3 per person. There are, however, marked differences among the municipalities in water use (Dar, 1996). In general, high-income communities consume more water per person than localities where the average income is low, as shown in Figure 14.3. Income may affect consumption in at least three different ways:

- High-income families have larger houses and gardens, cars to wash and some even have swimming pools to fill.
- Lower-income families are larger and obtain economies of scale in water use.
- Low income communities have poor distribution systems with more frequent interruptions in supply. This factor was important in the past but is less significant now that most municipal systems have been modernized.

The income elasticity of demand for residential water, estimated from the data depicted in Figure 14.3, is 0.60. This means that, on average, if income in community A is 10% higher than in B, residential water consumption in A will be 6% higher than in B. Figure 14.3 depicts data for a one-year cross section. Long period development is presented in Figure 14.2. In terms of Table 14.1, the household series in Figure 14.2 is urban water plus other and losses,[38] and it covers not only urban, but also assessments of residential consumption in the rural areas. Recall that the household series increased by 0.66% a year. To what extent was the long-term growth in consumption determined by changes in income? At this stage, the question can be answered only in rough terms. Per-capita GDP rose over the last 40 years at an average yearly rate of 2.64%. Accepting GDP as a proxy for income and applying the elasticity estimated from the data of Figure 14.3, the predicted growth in per-capita household water use is 1.58% per-year— much higher than the actual rate of 0.66% per year. Over forty years, an annual growth of 1.58% is an increase for the period of more than 85% compared to the actual increase of less than 30%. Evidently, income has not been the only determinant of water utilization in the urban sector in recent times.

[38] Unaccounted for water is the difference between water metered at the city gate and water sold to users.

Table14.1 Municipal Water Use in Israel, 2000

Sector	Per-capita (m^3)	Share (%)	Share (%)
Total Municipal	106.6	100	
Urban	86.1	80.8	100
Residential	60.5		70.2
Schools	3.4		4.0
Sport	1.1		1.2
Parks	7.2		8.4
Institutions	3.2		3.7
Hospitals	0.9		1.1
Hotels	2.1		2.4
Commerce	5.7		6.6
Police and Transportation	0.6		0.7
Construction	1.5		1.7
Industry	5.4	5.0	
Agriculture	3.1	2.9	
Other	1.4	1.3	
Losses	10.7	10.0	

Source: Dauber, Joseph, 2001, Annual Report 2000, Water Consumption in the Municipalities, The Water Commission, Tel Aviv (Hebrew).

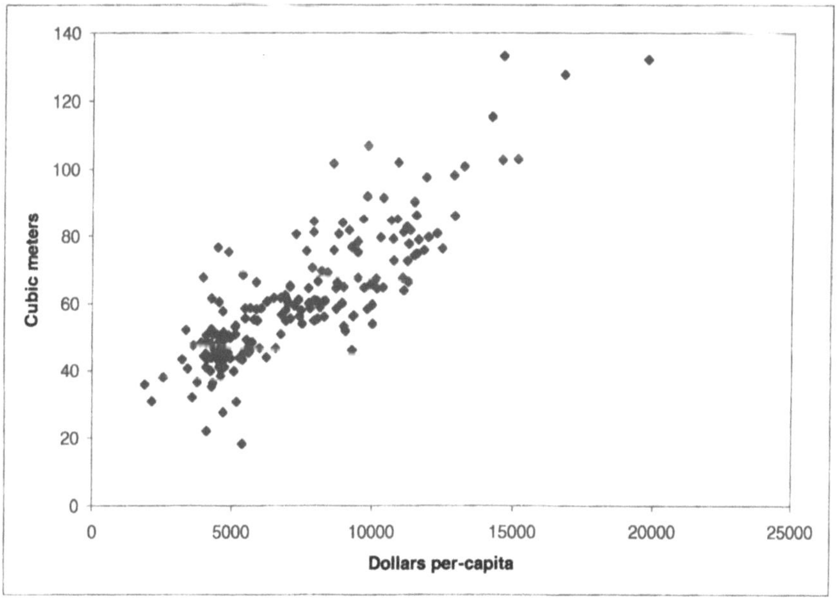

Figure 14.3 Income and Residential Water Consumption in Israel for 2000
Source: Cohen, Nir, M.Sc. Thesis (in preparation) The Hebrew University.

14.7 Prices and Cost

The government sets the prices for Mekorot's water, the charges users pay for water they receive from municipalities, and the extraction levies. Regional cooperatives charge their members cost-recovery fees, and wherever trade occurs, prices are market determined.

Prices should reflect cost. In the past, two major factors have affected the cost of water as conventionally defined (capital and operating cost). The first was the shift from relatively inexpensive to higher cost systems and the other was the rise in the world price of energy since 1974. In rough terms, and with the present prices of energy, water cost is \$0.12 per m^3 for local provision and the average cost is \$0.35 for countrywide supply through the National Carrier. Most of Mekorot's customers pay the same prices whatever the source of water and the prices they pay have risen over time, particularly when energy cost increased. The company is recovering approximately 80% of its cost. Israel is now entering the epoch of desalination with marginal water cost reaching \$0.60 per m^3 and perhaps even more.[39]

Another cost component, often not recognized explicitly, is the scarcity cost of water in its sources. When the quantity of water is limited, the supply to one person reduces the amount available to another. As an example, the reduction in production on the farm where water is missing is the scarcity or opportunity cost of water supplied to other users. This economic insight is the basis for the recently imposed extraction levy. The revenue of the levy goes to the general government budget on the presumption that just as the water is the property of the public at large so also payment for its use belongs to all the citizens of the country. Extraction charges, or pump fees, are imposed in many countries and sometimes for dedicated funds (OECD, 1999), but I have not found that the charges are connected, as attempted in Israel, to the scarcity cost of the water.

In the urban sector there are currently four different government-set prices and price groups for water:[40]

- The wholesale price the city pays Mekorot for water delivered at the head of the system (the city gate) is \$0.38 per m^3.
- An extraction levy (paid to the government) of \$0.13 per m^3 on water the city draws from its own wells (the levy on water for urban use drawn from surface sources is \$0.11 per m^3 and this rate applies only to the city of Tiberias that gets its water from the Sea of Galilee).
- The prices users pay the city for water. Residential consumers pay block rate charges in two-month billing periods:
  \$0.61 per m^3 for the first 16 m^3 per period
  \$0.89 for the next 14 m^3 per period
  \$1.27 per m^3 for any addition.

[39] The implicit price of the agreement signed recently with the builders and operators of the new desalination plant is \$0.53 per m^3, but to this will be added the cost of treatment and connection to the National Carrier.
[40] New users are also charged connection fees to the water and the sewage networks.

- Large families have discounts; a certain quota of garden water is also charged the lower rate ($0.61 per m^3).[41] In addition, households and other users pay $0.03 per m^3 to a restoration fund managed by the Ministry of the Interior.
- The fourth is the group of charges on other sectors in the urban economy. They pay different prices; for example, manufacturing charges are $0.36 and hotels $0.42 per m^3, both prices are flat rates.
- While water charges are identical in all towns and cities, municipalities may, with the approval of the Ministry, impose a sewage charge to finance investment in collection and treatment systems. These charges are not identical; they range from $0.20 to $0.60 per m^3 with garden water excluded.

Cities and towns pay a wholesale price of $0.38 per m^3 at the gates, which covers a little more than the cost of water of the National Carrier. Since agriculture pays less—approximately $0.25 per m^3 as average charge of a block rate schedule—it is generally claimed in Israel that water consumers in town subsidize farmers. This is not the complete picture. The cost of water Mekorot supplies to farmers in remote and hilly areas is higher than the price they pay. Likewise, the cost of delivering water to Jerusalem in the hills is twice as high as the cost of supplying water to Tel Aviv on the coast. With identical prices, some farmers and many town dwellers subsidize water used by other farmers and other city people. In addition to this cross subsidization, the government finances from the national budget the share of cost Mekorot does not recover with the administratively dictated prices.

14.8 Municipal Cost and Surplus

The municipality pays for water purchased from Mekorot and it incurs cost of pumping its own water (if it has any), distribution to users, and collecting and treating sewage. The prices and levies set or approved by the government are supposed to cover cost of service and maintenance of the system. It has often been claimed that local politicians—not unlike their colleagues at higher levels of government—are short run maximizers. The municipalities enjoy cash surpluses in their water operations and neglect maintenance while using the funds saved to finance visible, aboveground projects. The outcome is obsolete networks, water losses, and local authorities whose interest lies not in saving, but in increasing water use. Here are the words of a parliamentary committee (The Knesset, 2002): "The committee finds fault with water pricing in the urban sector. The gap between the price that the municipality pays for water and the price it charges the local consumer is not reasonable. The current situation can be defined as extortion." Nowhere in its 300-page report did the committee back its accusations with data or analysis.

Reality is a little more complicated. True, on average, cities and local councils have a financial surplus of 25% in their water and sewage service,[42] but there are

[41] The quota—applicable in the summer months of April to November—varies with the size of the garden, up to a limit of $30m^3$ per two-month billing period, provided the household's water consumption was higher than 16 m^3 per period. Residents in condominiums share the quota.

significant differences among the municipalities (CBS, 1998). Since the government dictates identical prices in all municipalities, those which are more efficient in both service provision and collecting water charges are making handsome "profits" on the water they supply, while others face difficult conditions such as old networks, hilly terrain and large families and are incurring losses. The significance of the overall surplus should also not be exaggerated; since it amounts to less than 7% of the property taxes the municipalities collect and 2.5% of income they have from all sources. Moreover, the municipalities invest regularly in development and restoration of water and sewage systems and their financial surplus may go to cover these outlays.

14.9 The Restoration Fund

Although a large part of the water system in Israel is relatively new, many municipalities also have older networks. Old pipelines tend to leak and to have calcified deposits on their walls. The deposits separate and pollute the water with abrupt changes of pressure or interruption and resumption of flows. To reduce leakage a fine has been imposed on cities and towns where water losses exceed 12% of supply. Twenty years ago, a Water Administration was established in the Ministry of the Interior. The Administration collects contributions to the restoration fund and assists cities and towns in investments and renewal of water networks. The participating municipalities receive assistance from the restoration fund and from other sources the administration raises. They have to match the external assistance with local funds. Participating municipalities also receive engineering and planning advice. Water losses have been reduced markedly since the establishment of the administration and the commencement of its activity; on averages the share of losses in supply is now 10% (Table 14.1).

14.10 Quality and Health

Utilization of water resources alters the quality of water stored in aquifers. When irrigating with aquifer water, salts percolate and their concentration in the aquifer gradually increases. Even if the accumulation of salt is slow—water sources are tolerant—eventually the aquifer will have to be cleaned, perhaps by desalination. The authorities and the public will have to recognize that maintaining sustainable resources and sustainable salt regimes is expensive but unavoidable. Sodium chloride (table salt) is the most common agent contaminating the aquifers, but it is not the only one. Among the other factors are nitrates from fertilizers and sewage, industrial chemicals, and fuel from gas stations. Encountering contaminated water, the Water Commissioner ordered wells to be closed, but this is not a solution as the contamination remains in the aquifer or even accumulates and spreads. Recently cities and towns in the coastal area have installed water-purifying devices and returned inactive wells to operation. The

[42] As is common in government budgets, the reported information is on current outlays, depreciation is not included. Cost of capital is included only as much as the municipalities pay interest on loans taken for the water and sewage sector.

Commissioner encourages the increased supply of water thus created and the government subsidizes the investments.

In a small country, where population density is high, water quality is always at risk. Diseases, particularly intestinal diseases, are a constant threat. By law, the responsibility for the hygienic standard of the water lies with the Ministry of Health. By the ministry's regulations, all water in Israel is chlorinated and large communities (45% of the population) receive fluoridated water. Suppliers, cities, and towns follow directives handed down from the ministry and the ministry also monitors water quality and takes action when and where its interference is needed. By a recent regulation, municipalities must publish their water quality tests. And indeed, tap water quality is generally good in Israel. The credit goes to the stringent application of the law by health authorities and to the concentration of large parts of the supply in one hand, Mekorot. With economies of scale and cost-plus pricing, the company incorporated in its network sophisticated quality monitoring, testing, and treatment facilities. Still, water quality assurance needs constant vigilance and effort.

Growth of urban population and its water consumption has entailed reductions in the supply of fresh water to agriculture and increased prices. At the same time, urban centers have increased the quantity of treated sewage that is mostly used for irrigation. Irrigation with recycled water is regulated to assure that contamination does not reach aquifers or wells and that the treated water does not come in direct contact with food. Modern practices allow farmers to mix chemical fertilizers with water in high-pressure drip irrigation systems (fertigation). The Health Ministry insists on separate networks and the fitting of check valves as an additional line of defense against unwanted chemicals in water supplied for human consumption. Similar check valves are also required in hospitals and manufacturing facilities. The ministry has been halting the development of residential areas unless adequate sewage removal was assured.

Surface water poses special health hazards. Small localities in the north, drawing water from the River Jordan or the Sea of Galilee, are required to filter the water they supply to households. Sea of Galilee water supplied by the National Carrier is not up to most stringent standards; particularly, turbidity cannot be avoided. The technical solution is to filter the water in the National Carrier. The Treasury deemed the solution too expensive and farmers opposed it, fearing that they, who are also using the carrier's water, will be required to pay the cost of filtering that they actually do not need. An alternative solution is to filter the water at the gates of the cities. Economically a single filter is more efficient and it secures quality to all users, including inhabitants of small communities that may find local filters too expensive or technically too complicated to operate. The filter has not been installed as yet, but preparations are underway to fulfill the government's promised construction of the filtering plant. The government was pushed into action when a green group recently brought the issue before the Supreme Court.

Indeed, while green lobbies have not been effective in changing overall water policies and reducing overdrafting, they are heard and are often effective on issues of water quality. They bring the authorities to court whenever they find that water is not up to the standards specified in health regulation, in Israel or in other countries. They also increase public awareness of water quality issues. They may thus be partly responsible for the increasing use of fashionable bottled "mineral water" in Israel. Evidently consumers in a high-income country are willing to pay 500 times the price of water supplied in the national network for bottled drinking water.

14.11 Sewage Collection and Treatment

Fifty to sixty percent of the water supplied to urban communities is returned as sewage. Most of the water households and others in towns and cities receive is not consumed but is used as a carrier of dirt and refuse. Once treated, the sewage—now as recycled or reclaimed water—can be reused. Close to a third of the water provided to agriculture in the year 2001 was recycled sewage. This share is expected to exceed 70% in twenty years.

Originally, houses in Israel had individual septic tanks and cesspools in which the sewage was collected and from where it slowly drained away, often into aquifers. When collection systems were first built, some 50 years ago, the raw sewage was dumped directly into the Mediterranean Sea or the nearest streambeds going to the sea. This practice created an environmental nuisance and health hazards. Most visible was the pollution in Tel Aviv where waves washed parts of the sewage back on the beach. In the mid 1970s, the city, together with a few surrendering towns and with the help of the World Bank[43], built a central treatment plant (the Shafdan, named after the biblical term, Dan, for the region in which Tel Aviv is located).

Regulations on sewage removal were on the books for a long time and a comprehensive law requiring modern treatment was enacted in 1962. The requirement was only partly satisfied. As Table 14.2 indicates, 30 years ago less than 40% of the sewage was treated. But the situation has gradually improved. The sharp increase in the treated sewage reported in Table 14.2 between 1971 and 1982 is due to the construction of the Shafdan, the Tel Aviv treatment plant. However, it and another plant built in Haifa by Mekorot were essentially the only high standard treatment plants. Most other municipalities, if they treated their sewage, treated it superficially and did not bring the recycled water up to the required standards. Evidently, as with water systems, city fathers preferred to invest available funds in visible projects and let partially reclaimed sewage flow away down the nearest creek. The government, as much as it attempted, did not manage to force compliance with the law and with health regulations.

The policy was modified dramatically in 1992 when a new government came into office, replacing the stick with the carrot.[44] The national government started to enthusiastically support construction of local sewage treatment plants. Budgets set for assistance to municipalities for sewage collection and treatment increased fivefold between 1991 and 1995, and the money has since continued to flow in similar rates. This change came in the wake of the realization that the sewage problem and the extreme negative externalities it created would not be solved unless the local authorities were assisted. But two other considerations also added to the move to the new policy.

---

[43] This was the last project the Bank undertook in Israel. The country has since matured to the status of a developed economy.

[44] On the question of fines or subsidies as policy instruments, see Finkelshtain and Kislev (1995).

One was a wave of immigration increasing significantly the demand for housing and employment and the second was a shift away of the new government (in 1992) from investment in settlements in the West Bank and Gaza Strip to infrastructure inside Israel.

Table14.2 Sewage and its Treatment in Israel

Year	Raw sewage	Centrally Collected		Treated	
	Million m^3	Million m^3	Percent	Million m^3	Percent
1971	183	145	78%	69	38%
1982	251	210	84%	137	55%
1989	293	273	93%	232	79%
2000	451	438	97%	418	93%

Source: State Comptroller, *Annual Report 41* and data received from the Water Commission.

Some of the newly constructed plants treat sewage that was disposed raw in the past, but most replace or modernize old and primitive facilities. The recycled water is of secondary treatment (removal of organic material), except for the Tel Aviv plant where the treatment includes filtering in sandy dunes and the recycled water is considered "fit for occasional drinking." To reduce the financial burden of the municipalities and assure professional management, the cities and towns are encouraged to contract with business enterprises to Build Operate and Transfer (BOT) the plants. The plants transfer to the municipality after a period, typically, of 20 or 25 years. The introduction of private interests into the provision of local public goods, in swage treatment plants and in the future perhaps also through the incorporated water units, raises issues of finance and control that municipalities in Israel are now experiencing for the first time.

Most recycled water is utilized in agriculture where farmers have accepted it as the provision of fresh water has been curtailed and its price increased. At present, recycled water supplies 30% of the water delivered to agriculture. This share is expected to grow as the supply of fresh water is increasingly diverted for urban sector use and the urban sector produces larger and larger quantities of sewage for treatment and disposal. The government provides most of the investment funds for the adaptation of irrigation to reclaimed water. The cost of the adaptation is not negligible: storage is prepared to keep treated water from winter to summer and new networks are constructed to keep recycled sewage separate from drinking water. Farmers and their regional coops cannot raise on their own the amounts needed for these projects on the capital market.

Up to now, most experts in the water economy have regarded secondary treatment and utilization in agriculture as an appropriate solution to the sewage problem. There is however, a growing understanding that the treatment will have to be more thorough to reduce health hazards and damage to the soil. Some specialists (in particular Zaslavsky, 2002) have recently been calling for desalination of all reclaimed water. The arguments brought in support of desalination are:

- It will reduce the amount of salt irrigation adds to the aquifers (households and industry add salt to the water they use and dump as sewage)
- Recycled water will be upgraded to tap-water quality and there will thus be no need to have separate networks for the reclaimed water
- Desalinated sewage could be used for artificial recharging of the aquifers, saving in this way water that is evaporated when reclaimed sewage is stored in open-air reservoirs.

The sewage desalination idea, which may increase significantly the cost of treatment that city dwellers will have to cover, has not been accepted thus far by the authorities but it may be considered seriously when actions to maintain sustainable salt regime in the aquifers become real.

14.12 Incorporation

Local governments are not necessarily efficient suppliers of services. A newly enacted law empowers the municipalities—individually or in town associations—to establish business entities, corporations that will function as public utilities supplying water and taking care of the sewage. But the law does not only empower, it permits the Minister of the Interior to establish such a utility for a hesitant municipality. The envisioned institutional structure is made of four layers:

- The municipality sets-up the utility and transfers the city's water department to the new business enterprise.
- An Officer in Charge in the Ministry of the Interior will approve the utility or close it down if it fails to provide the expected services.
- A national Water Public Utility Authority will set standards and prices for the utilities and monitor their operation.
- The utility will receive from the municipality all assets and facilities of the water department together with its obligations, bank accounts, etc. It will be responsible for the provision of services in the water and sewage area.

The water utility may be owned by the municipality, it may be an investor-owned utility or its ownership may be shared by the municipality and private interests. The law explicitly specifies that the major supplier of water to a community cannot operate the local water utility. Therefore, Mekorot cannot operate as a local utility. This law was enacted in 2001 and the new structure is still in its teething stage so there will be learning by doing. A pioneering experiment in the incorporation direction is being conducted in the capitol city of Jerusalem.

Jerusalem straddles a mountain range; the largest part of the city is drained to the west toward the Mediterranean and a smaller area is drained to the east, to the Jordan River and the Dead Sea. For decades, the city dumped its raw sewage into creeks on both sides. This was exceptionally irresponsible behavior as the sewage was disposed above the mountain aquifer that supplies the best quality water in Israel. Several years ago the city, with the encouragement and financial support of the government, established a water and sewage utility and the utility constructed a treatment plant on the western outskirts of the capital. Instead of raw sewage, recycled water now flows down the creek. (The farmers downhill are complaining that the treatment has reduced nitrogen concentration in the water.)

All workers of Jerusalem's water department moved to the new utility when it was established while maintaining their previous pay and working conditions. Gradually, as workers retire, the utility reduces its work force and trims its operation. A reduction in water charges is envisioned. As of now the utility is wholly owned by the city, but privatization is planned. Jerusalem's experience is being watched closely by other municipalities that have to weigh the loss of freedom to run their water economies, sometimes profitably, against efficiency and quality of service.

14.13 Water Policy

Municipalities are responsible, in many places, for the water they supply to users in their jurisdictions. Not so in Israel. The country's Water Law made all water the property of the public, managed by the state. Mekorot's main lines reach every locality and, even municipalities that own and operate wells may purchase unlimited amounts of water from the national company. Consequently, the management of the natural resources and questions of sustainability have not been relevant at the municipal level in modern times. Similarly, municipalities readily dumped sewage in streambeds, particularly if it could then flow quickly away. Jerusalem was a blatant example.

The national government now supports the local water and sewage sector handsomely and its money cures many deficiencies. The urban water and sewage systems have been and are being upgraded, and it can safely be asserted that by 2010, almost all communities in Israel will have modern water and sewage facilities and the recycled water will be utilized in agriculture according to health and water safety regulations. Needless to say, quality will still have to be guarded, criticism will continue to be heard, and not all professional debates will be settled.

The government, encouraging and supporting the modernization of local water and sewage systems, may be seen as a benevolent sovereign, concerned for the welfare of its citizens. However, the financial support of the water economy is but one aspect of the involvement of the central government in the affairs of local authorities. Such intensive interference suppresses local participatory democracy and reduces the willingness of the public to try and influence in democratic ways the society they are part of. It need not be: the urban water sector can be a lever raising local democracy as suggested in the following two examples.

Recall the charge that municipalities are making profits on their water service. In principle, the financial surplus in the water sector, wherever it exists, can be viewed as a tax levied by the local authority on a service it provides and as a subsidy where losses occur. As the law does not specify any such tax, its implicit imposition can be and should be criticized. But what should an efficient municipality do? Dump the surplus money down the drain? Moreover, water service is a convenient tax base; collection is easy and effectively costless. Municipalities should be allowed to levy a water tax and perhaps also other similar taxes. If a water tax were allowed, it would be imposed explicitly by a decision of the city council or a referendum. If the voters do not like a city-imposed tax, they will vote the mayor and the council out, come next election. Imposing taxes and using the revenues for community projects would strengthen local democracy and improve its functioning.

The second example also relates to prices. Wholesale water prices, at the city gate, reflect actual cost of water supplied from natural sources. In two or three years, the country will be entering the desalination epoch. The declared policy is that future wholesale prices will be equal to realized average cost and will not be much affected by the desalination of relatively small quantities of water. Average cost pricing with small increases in price as desalination expands will probably be accepted by the public. Such prices seem just and do not have much effect on the small budgets households spend for water. Economically, however, they are not necessarily the right prices.

By the assessment of the Water Commissioner, founded on government's decision on allocation to agriculture, the supply from the natural resources no longer covers the water needs of Israel. If this shortage justifies desalination, the monetary value of the water (its scarcity cost) is equal to at least the cost of the new source of supply--approximately \$0.60 per m^3, more than 50% higher than the current price the cities pay. The wholesale water prices could be raised now to \$0.60 per m^3, thereby imposing the economically efficient marginal cost. User prices in the municipalities would also go up in accordance with this change. A sudden increase in prices will intensify awareness of water shortage and reduce consumption when supply is tight. In the long run, such marginal cost pricing will encourage the authorities to seek local solutions to water supply problems. Tel Aviv or Haifa may contract privately with desalination companies, reducing in this way their reliance on the national system and the monopoly that operates it.

Naturally, city residents would not willingly accept this large a price increase. Surprisingly, the Treasury experts have also rejected the idea of marginal cost pricing. They want to see identical farm and urban wholesale prices and they realize that a price of \$0.60 per m^3 for fresh water supplied to agriculture will reduce the quantity demanded, and therefore irrigated areas and farming activities, to a level far below the intention of the government. If the decision to maintain a certain level of irrigated farming is maintained, water supplied to agriculture will have to be subsidized and prices in town will be higher than for irrigation. Urban water prices may as well be based on marginal cost.

14.14 Structural Reforms

Several government ministries and agencies are responsible for water issues, the Water Commissioner being only one of them. The recent water crisis has intensified the calls often heard in Israel for reorganization and that all responsibility be concentrated in one hand. The general issue is outside the scope of the chapter but multiple authority can also be seen in the government bodies dealing with urban water: The Water Administration in the Ministry of the Interior is responsible for restoration of water networks, a Sewerage Administration in the Ministry of Infrastructure deals with sewage lines and treatment plants, and a separate office under the Water Commissioner assists the farmers in the utilization of recycled water. This situation may create inefficiencies; for example, when treatment plants are not coordinated with irrigation possibilities. Concentration of government functions in the urban water and sewage sector in one hand could streamline the administration and increase its efficacy.

The "incorporation" is aimed at creating business-like enterprises in the urban water sector. An effort of similar nature is going on with respect to Mekorot. For years the government has been trying to divide the company into several sub-units and to increase its capacity to operate independently in the capital market, away from the government's purse. To reduce Mekorot's monopolistic position, the government also decided that the company cannot build desalination plants and a "wheeling" clause has been added to the Water Law by which Mekorot (or any supplier) is required to move water through its network for "third parties," from another supplier to its customers. Mekorot is also prevented from becoming a local water utility under the incorporation law.

Among the government agencies, the Treasury has been at the forefront of the drive for incorporation, the attempts to restructure Mekorot, and the introduction of business enterprises into the area of sewage treatment, but its efforts have not always been appreciated. The parliamentary commission (The Knesset, 2002), as one example, depicted the Treasury as arbitrarily halting the development of new projects to the detriment of water supply. It can be expected that structural changes, experimentation, and public debates will continue for a long time to come.

References

Allan, T., 2000. The Middle East Water Question. Tauris, London.

Auerbach, O. and O. Karassin, 2002. "Management of Water Resources in Israel," in: *Paths to Sustainability*, Shadow Report to the Government of Israel's Assessment of Progress in Implementing Agenda 21, presented at the World Summit for Sustainable Development, Johannesburg, www.worldsummit2002.org/?regions/arabmiddleeast/israel.

CBS (Central Bureau of Statistics), 1998. *Local Authorities in Israel, 1995 Data* (Hebrew).

Darr, P., 1996. *Factors Affecting Urban Water Consumption* (Hebrew, mimeo).

Finkelshtain, Israel and Yoav Kislev, 1995. "Prices vs. Quantities: The Political Perspective," *Journal of Political Economy*, 105:83-100.

Hardin, G., 1968. "The Tragedy of the Commons," Science, 162 (3859):1243-1248.

Israel Water Law—1959, (translated from Hebrew) unknown translator and publisher.

Karlinsky, N., 2000. *Citrus Blossoms, Jewish Entrepreneurship in Palestine, 1890-1939*, Magnes Press, (Hebrew).

Kislev, Y. (forthcoming). "The Water Economy of Israel," in: Rageb, Jamil, David Hamberg, and Joseph Ginat (eds.) *Water in the Jordan Valley*, Oklahoma University Press.

Kliot, N., 1994. *Water Resources and Conflicts in the Middle East*, Routledge.

The Knesset (parliament), 2002. *The Report of the Parliamentary Inquiry Committee on the Water Sector* (Hebrew).

OECD (Organization for Economic Cooperation and Development), 1999. *The Price of Water—Trends in OECD Countries.*

Water Commission, 2000. *The Tasks of the Water Economy of Israel in the Long Run* (Hebrew).

Zaslavsky, D., 2002. *Under the Red Line—On the Water Crisis in Israel*, the Technion, Israel Institute of Technology.

Zusman, P., 1997. "Informational imperfections in water resource systems and the political economy of water supply and pricing in Israel," in: Parker, D. D. & Tsur, Y. (eds.) *Decentralization and Coordination of Water Resource Management*, Kluwer Academic Press, Boston.

Chapter 15

The British Experience

Kenneth F. Clarke
Water engineering and management, United Kingdom

15.1 The Geography of the United Kingdom

The United Kingdom is comprised of four political units: England, Scotland, Wales, and Northern Ireland, each with a different geography and culture (Encyclopaedia Britannica, 1996). The United Kingdom possesses a long history, and strong political and cultural links with other areas of the world. The political system of the United Kingdom is characterized as stabile and consistent. Furthermore, the United Kingdom is part of the European Community.

The hydrogeology is diverse. The principle aquifers, situated in the middle, the south and the east of the country, are the Chalk, the sandstones and pebble beds of the Bunter, and the sandstones of the Coal Measures. The topography, the geology and the weather are the main factors that create the following phenomena which dominate how water resources were developed.

- The highest rainfall is in the north and west of the Kingdom and the lowest in the south and east. However in broad terms the population is the inverse as it is higher in the south and east, with London being the major centre, and low in the north and west. The south and east have in the past been mainly supplied with water from local aquifers. However these resources have become fully utilized which means in general that water has to be moved from the north and west of the country to the south and east.

D.E. Agthe et al. (eds.), Managing Urban Water Supply, 241–255.

- The mountains are hard and impervious and therefore there is no groundwater and the water sources in the rivers have been supplemented by the construction of many reservoirs. Because of the higher rainfall in the west and north, these reservoirs have to be designed to withstand a drought which lasts little more than a year.
- In the drier south and east there are good aquifers which are now fully utilized. Reservoirs have to be designed to withstand droughts of three years or more.

15.2 The History of Water Supply and Sewage Disposal in the U.K.

In pre-Roman times there was plenty of water for the comparatively small population. Water was taken from springs, ponds, wells, streams and rivers. The importance of water is illustrated by many of the surviving place names such as Oxford. It was not only a crossing place of the River Thames but also a place where the people could get water for themselves and their animals.

The growth of urban communities created the need to channel water supplies from more distant sources of water and therefore the Romans built aqueducts. Channels were constructed of cut stone, brick, rubble, or rough concrete. Pipes were made of drilled stone or of hollowed wooden logs; clay and lead pipes were also used. Perhaps the most important use of lead was for pipes to supply fresh water to buildings. The Romans provided generous water supplies for their cities; all of the supply systems worked by gravity and many of them used aqueducts and syphons. Although most people had to carry their water from public fountains, there was limited distribution of water to public baths and some private residences. Private and semiprivate baths and latrines became fairly common. The water supply systems that the Romans had developed were largely destroyed by barbarian invaders after the fall of Rome (King, 1992)

In 1236, the Corporation of the City of London obtained a royal grant to convey water from the Tyburn stream to the City in six inch (15cm) lead pipes. In 1439 the Mayor of London asked the Abbot of Westminster to supplement the water supply to his 50,000 citizens. In the 16th century when the smaller religious communities were suppressed, local cities took over their water supply systems (King, 1992).

Early water supply systems diverted water, which flowed by gravity, from the source into the towns. The water usually flowed into a large tank and then through taps or spouts into stone basins. The people collected their water from these basins themselves, or had it delivered to them by water carriers or horse drawn water carts. However this method of carrying water was cumbersome, slow and unreliable and by the 17th century the carriers could not keep up with demand. Therefore various pumping devices were tried. Boulton and Watt steam engines were first installed to pump water in London in 1787 (Encyclopaedia Britannica, 1996).

At the same time as these developments in the motive power for pumping, there were also developments to improve the quality of the water. In 1791, James Peacock took out a patent for a process of water filtration which included the ability to clean the filter by reversing the flow; a principle still used in modern day filters. Other successful filters were designed and installed by John Gibb (1804), Robert Thom (1827) and James Simpson (1827) (Institution of Water Engineers, 1961).

Water provision was undertaken by local municipalities and small local private companies. They worked conscientiously under the effective control of the Ministry of Health using legislation contained in the Waterworks Clauses Acts of 1847 and 1863 (Institution of Water Engineers, 1961). Their powers were extended by the Public Health Act 1936. So effective were they that, apart from this legislation, there was no need for further legislation until the Water Act 1945. Under this Act there commenced the process, which is still ongoing, for the amalgamation of water undertakings, whether they were local government or private companies, by the creation of larger local government organizations or water boards (Institution of Water Engineers,1961).

By the late 1950s the increasing population and its increasing per capita consumption meant there was a growing demand for water. This put pressure on the rural community for the use of agricultural land for reservoirs and the use of ground water which dried up streams which had been traditionally used for agriculture. These demands also put pressures on the environment not only directly but also indirectly because the higher water consumption meant higher levels of sewage in the water courses. Therefore the government of the day decided that there was a need in England and Wales for a more nationally planned approach to the development of water resources. As a result the Water Resources Act 1963 was passed by parliament. Under this Act, 26 River Authorities were set up and they had a duty to plan water resource development on their respective river catchments (Twort, et al., 1994). One of their key functions was the licensing of all water abstractions, apart from single households abstracting from aquifers. The licences were granted on what the abstraction pumps were capable of pumping and not on what the water resource could stand. No account was taken of all uses of the resource including the environment. To coordinate the Water Authorities, the Water Resources Act 1963 set up the Water Resources Board to plan water resources development nationally. It did this coordination by setting up joint working parties in three key regions of England and Wales. However the Board did not have the powers to undertake any regional water resource development schemes. Other weaknesses of the Water Resources Act were that the Water Authorities did not have the powers to pay for the re-siting of sewage outfalls or to improve the effluents of those discharges, nor did they have the powers to redistribute more equitably water resource abstraction licences.

These failings led to an extensive reorganisation of the water industry as promulgated by the Water Act 1973. Under this Act the responsibilities of the River Authorities and the Water Resources Board, and all the water supply undertakings and the sewerage organisations owned by local government were passed to 10 regional water authorities (Twort, et al., 1994). Those private water companies that were in existence in 1973 were kept in existence because the government of the day did not want to be seen to be nationalizing them. The area of each authority covered several river catchments and they were the first organisations with the responsibility for total river catchment planning and management. They provided the services of water resource planning and management, water supply, sewerage and sewage treatment, land drainage and flood prevention, river water quality, and fisheries. These new authorities were effective and with their large financial and technical resources they improved the quality of the water, sewerage, and sewage treatment services and were responsible for a major improvement of the quality of water in the rivers.

The government of the 1980s had the strong conviction that: private services were more efficient than public services; the public sector should be reduced; water services should not be financed with public money; and private companies would have the freedom to create more commercial opportunities both at home and abroad (Twort, et al., 1994). Although the water authorities had been successful, the government changed them into Public Limited Companies in 1989 (i.e. they privatised them).

Some functions of the water authorities, however, could not be privatised because the authorities had many duties, such as the licensing of water abstractions, which could not be put under the control of a private company (Twort, et al., 1994). Therefore the duties of the water authorities were divided between the new privatised companies and a new organisation, the National River Authority [NRA], which was later to become, with others, the Environment Agency. The new companies became responsible for the water and sewerage services while the NRA took over the duties of: licensing water abstractions and effluent discharges, regulating fisheries, controlling navigation on rivers, managing land drainage and flood prevention and improving flood defences.

15.3 Water Supply Models

There are many types of organisations which supply water throughout the world. Some examples are:

- The infrastructure of water supply is owned by a commercial company which is also responsible for its operation.
- Commercial firms build water supply facilities at their own expense, operate those facilities for a specified length of time and then transfer the responsibility to a governmental unit. These schemes are known as BOT which is an acronym for Build, Own and Transfer.
- The local government maintains ownership of the water supply and sewage treatment infrastructure and a private company operates the infrastructure, some times being responsible for investment and construction to increase the capacity of the infrastructure.
- A private company is responsible for the operation and maintenance of the infrastructure but the assets are still owned by the local government which also carries out any increases to capacity and bills the customers.
- Larger companies give support to the managers of a water supply organisation on a partnership or consultancy basis.

15.3.1 ENGLAND AND WALES

In England and Wales, the water supply and sewage services are run by Private Limited Liability Companies (PLCs) which own the assets of the infrastructure. To make their shares more attractive when they were set up in 1989 they were empowered to engage in commercial activity (Twort, et al., 1994). Their water and sewage functions were put under wholly owned but separate companies known as the PLC's core business. Using this model the Government, through its regulators, keeps control over the core business of each PLC but still allow the PLCs to engage in commercial activities of their own choosing.

15.3.2 SCOTLAND

In Scotland, the water supply organisation model chosen is different to England and Wales. The Scots were not prepared to go along the privatisation route adopted by their larger southern neighbour being concerned about the commercialisation of a natural monopoly. Therefore, Scottish Water [SW] was created in response to The Water Industry (Scotland) Act 2002 using a public sector model in that SW remains answerable to the Scottish Parliament. However, it is structured and managed like a private company, with 12 Board members, comprising five executive directors and seven non-executive directors. SW is a new organisation replacing East of Scotland Water, North of Scotland Water and West Scottish Water. It provides water and waste water services to household and business customers across one third of the land area of the United Kingdom. It serves over five million customers in Scotland with approximately 2,500 million litres of water a day and also provides them with sewerage services (Scottish Water, 2002). SW is the fourth largest water services provider in the U.K. and the 12th biggest business by turnover in Scotland. It has assets valued at £13 billion, including 441 water treatment works, 643 waste water treatment works and 1,127 service reservoirs (Scottish Water, 2002). SW is accountable to a number of Scottish regulatory bodies: the Water Industry Commissioner for Scotland; the Scottish Environment Protection Agency; and the Drinking Water Quality Regulator for Scotland (Scottish Water, 2002).

The Water Industry (Scotland) Act 2002 resulted from a detailed review of how the water industry was positioned to respond to the need for the biggest capital investment programme in recent history (Scottish Water, 2002). The Scottish Executive of the Scottish Parliament proposed that a single authority was better placed than three separate authorities to harmonise charges across Scotland to finance a huge capital investment programme; become a more competitive force in the public industry; and provide better value for money to customers as a result of economies of scale. So SW was created to enable the water industry in Scotland to keep charges at an acceptable level for customers.

15.3.3 NORTHERN IRELAND

In Northern Ireland there is yet again another model chosen by the Government to reflect the special political environment there. The Northern Ireland Water Service is an Executive Agency within the Department for Regional Development, a department which is also responsible for the roads services (Northern Ireland Water Service, 2002). It provides water and sewerage services to over 720,000 domestic, agricultural, commercial and business customers throughout Northern Ireland. It has an annual budget of £255 million and fixed assets valued at £4.9 billion. One of its strategic aims is to improve drinking water quality at the customer's taps, achieving 99% compliance with the Drinking Water standards set in the NI Water Quality Regulations by March 2006. Its other goal is to improve the quality of discharges from wastewater treatment works, in line with the Urban Waste Water Treatment Regulations (NI) 1995, achieving 80% compliance with wastewater treatment works discharge standards set by the Environment and Heritage Service by 2005 (Northern Ireland Water Service, 2002).

Over 99% of households in Northern Ireland have a public water supply and each day the NIWS supplies some 720 million litres of high quality drinking water to its customers. It maintains and operates a water infrastructure consisting of 48 impounding reservoirs, 47 major water treatment works, 490 service reservoirs, 270 water pumping stations, and 22,250 kilometres of water mains (Northern Ireland Water Service, 2002). Approximately 83% of households are served by the public sewerage system which collects, treats and disposes of about 126 million cubic metres of wastewater each year. It is regulated by the Northern Ireland Environment and Heritage Service which is under the control of a different government department (Northern Ireland Water Service, 2002). When direct rule was reintroduced in late 2002, the only change made to the management of NWS was that the minister in charge of the sponsoring department, the Department for Regional Development, became a United Kingdom minister instead of being from the Northern Ireland Assembly.

15.3.4 EVALUATION OF ALTERNATIVE MODELS

The model chosen by England and Wales has been an undoubted success and the private sector has demonstrated its ability to deliver within the water sector. Whether it would be a success in other countries with different cultures and geographies, different political systems, and most importantly differing extents to which the indigenous water supply organisations have already been developed is a matter for the countries concerned. However, one matter is clear: although there can be global strategies, water supply is essentially a local service, using local resources and a locally fixed infrastructure.

The success in England and Wales has been driven by two motivators. Firstly, there is the drive given by commercial enterprise and financial acumen. Secondly, there is the drive given by the three main regulators within England and Wales. The success of regulation in England and Wales is ensured by the almost total transparency in what regulation does: the setting of clear targets for the private companies, clear measurement of the key measures that demonstrate performance against targets, and the full reporting of the performance against the targets.

In 2002, it is too early to appraise whether the new model for Scotland is a success or not. It is probable that the model in Northern Ireland will change when it is seen there the tremendous energies and opportunities that have been created by the English and Welsh model. However Northern Ireland is unlikely to adopt the English and Welsh model and will probably adopt the Scottish model.

15.4 Regulation

When the water industry was privatised in England and Wales the government had in effect set up monopolistic water companies. Therefore it considered that there was a need to put in place a set of regulators to ensure that the asset bases of the new companies were not run down in order to make increased profit, the quality of water provided was good and the environment was protected (Twort, et al., 1994). Therefore it set up an economic regulator, a water quality regulator and an environmental regulator. Although there are different organisations responsible for regulation in Scotland and Northern Ireland, their roles are basically the same as for England and Wales.

15.4.1 OFFICE OF WATER SERVICES

The Office of Water Services, known as Ofwat, is responsible for making sure that the water and sewerage companies in England and Wales give good-quality, efficient service at a fair price. It is a government department led by the Director General of Water Services and it is the economic regulator of the water industry in England and Wales (Office of Water, 2002). This means that it limits the amount companies can charge customers; it makes sure that companies carry out their responsibilities under the Water Industry Act 1991; it protects the standard of service customers receive; it encourages companies to be more efficient; and it encourages competition between the companies where appropriate. One way it engenders competition is to compare the activities of all the companies and then publish the results. This encourages poor performers to rise to the standards of the best. The activities of Ofwat have four parts (Office of Water, 2002):

- Ofwat sets a limit on the yearly change in each company's average water bills although bills for individual customers can change by more or less than the average. Price limits, for the five years 2000-01 to 2004-05, came into force on 1 April 2000. On average this will reduce prices by 2.1% each year. This does not include the effects of inflation. Each year, Ofwat checks that companies are charging within the limits and that individual charges are fair to different types of customers: metered or unmetered, large or small, and in towns or in the country. Ofwat expects that charges should broadly relate to the costs of providing the service. So, whether or not the customer has a meter, his bill should cover the cost of supplying him with clean water, getting rid of his dirty water, and draining water from around his home.
- Ofwat also has a role in ensuring that the companies are economical and efficient because they are collectively investing over £8 million a day to improve drinking water and to protect the environment in England and Wales. Ofwat checks how the companies perform to make sure that customers get good value for their money. It expects companies to improve their services by being more efficient, not just by raising their prices. Companies have targets to reduce the amount of water that is lost through leaks from their networks. Ofwat checks that they are also encouraging customers to use water wisely. All companies now offer a free supply-pipe [i.e. the

pipe from the mains into each house] repair service.

- Ofwat has a role in protecting customers. It approves companies' codes of practice and provides guidelines to help them deliver better service to customers. It also checks that companies meet their responsibilities to customers, such as reducing the number of properties at risk of being flooded with sewage. Each year, it publishes information about how companies perform, and it takes action against companies that appear to be operating at a poor level.
- Ofwat also ensures standards of water supply and associated activities by promoting a Guaranteed Standards Scheme, which is a set of legally-binding standards. If a water company does not meet them, customers are entitled to a compensation payment. The Guaranteed Standards cover: making and keeping appointments within agreed times; answering customer questions about their accounts and their written complaints within agreed times; warning customers about planned supply interruptions that may last more than four hours and notifying them when service will be restored; payments they must make if a customer's home is flooded from their sewer; and payments they must make if they fail to maintain a minimum water pressure. Ofwat periodically reviews the Guaranteed Standards Scheme to make sure it reflects customers' needs. Some companies have compensation schemes that go beyond the minimum standards.

To ensure that the views of the companies' customers are clearly heard, Ofwat has set up 10 regional Customer Service Committees (CSCs). They are independent of the water companies and have their own legal duties, which are to represent the customers' interests, deal with complaints about water companies, and monitor the service the water companies provide. The CSCs also tell Ofwat about important issues that affect customers. They work closely with Ofwat and hold their meetings in public. CSC members are local people with a wide range of backgrounds, who are appointed by Ofwat to represent everyone's interests. A small team of full-time staff supports each committee (Office of Water, 2002).

To further protect customers Ofwat set up a National Customer Council (ONCC) which represents their water interests at national and European levels. The chairman of the ONCC and its members are all chairmen of regional CSCs. The ONCC advises the Director of Ofwat on customers' interests, helping him develop customer-friendly policies that meet his duties to protect customers' interests. The ONCC also aims to make customers' voices heard in Europe before agreement is reached on new European directives which may affect their water bills and service. If a customer is unhappy with his water company's service he can use the water companies' complaints procedure which is a legal requirement for them to have (Office of Water, 2002).

15.4.2 DRINKING WATER INSPECTORATE

The responsibility for regulating the quality of drinking water that is that of the Drinking Water Inspectorate [DWI] which is an independent organisation which oversees the quality of tap water (Drinking Water Inspectorate, 2002). It is a team of professionals with a wide range of experience and expert knowledge in all aspects of water supply, including chemistry, microbiology and engineering. On health matters, the team gets advice from the Government's Chief Medical Officer. The Inspectorate's main job is to check that the water companies in England and Wales supply water that is safe to drink and meets the standards set in the Water Quality Regulations. Inspectors of the DWI carry out technical audits of each water company which include an annual assessment of the quality of drinking water supplied by the company and inspections of each individual company. The assessment of water quality is based on information received regularly from the water companies, which includes the results of the millions of tests made each year to see if the water meets the standards. The team also carries out inspections to be sure that the reported test results are reliable and give a true picture of the quality of the water supplied. The inspections may cover many things such as checks that: the sampling procedures are satisfactory; samples are tested by trained staff using accurate methods; the right numbers of tests are carried out; correct results are entered in the public record; appropriate water treatment processes are used; the water distribution systems are operated and maintained satisfactorily; and customer complaints are handled properly. If they find something apparently wrong, they take action to ensure that the company puts things right (Drinking Water Directorate, 2002).

The Kingdom's Government has set legal standards for drinking water in the Water Quality Regulations. Most of these come directly from an obligatory European Community Directive but some United Kingdom standards are more stringent. Most standards are based on World Health Organisation guidelines. The standards generally include wide safety margins. There are standards for bacteria; chemicals such as nitrates and pesticides; metals such as lead; and the way water looks and how it tastes. If there are failures to meet the standards, the DWI assesses each failure: some are trivial or unlikely to happen again, so no further action is needed. Other failures will be put right by improvements already being carried out by water companies which may involve upgrading of treatment works and distribution systems, and together cost many hundreds of millions of pounds. Any remaining failures will be addressed by DWI requiring the water company to carry out further improvements quickly. The DWI has a duty to require water companies to take any necessary action and it checks that improvements are being completed on time (Drinking Water Inspectorate, 2002). The DWI also investigates complaints from customers and incidents which affect or could affect drinking water quality. Their investigations of incidents can lead to water companies being prosecuted.

Water companies are required to give local authorities information about the quality of water supply in their areas in their Annual Reports. These are lengthy, containing a great deal of detailed information about the quality of drinking water in England and Wales. They may be in a main library or at the customer's water company's offices. Companies also publish a free leaflet each year giving the results of the water company's monitoring (Drinking Water Directorate, 2002).

To ensure that the customers have access to a comprehensive range of information on water supply the DWI also prepares and distributes leaflets covering: 'How good is drinking water,' 'Chlorine, smell, taste discoloured water,' 'Lead in drinking water,' 'Private water supplies,' 'Problems with a customer's drinking water,' 'Water hardness,' 'DWI summary reports on each company's tap water,' 'Where water comes from and how is it made safe to drink,' and 'Using lead-free solder for drinking water fittings' (Drinking Water Directorate, 2002)

15.4.3 ENVIRONMENT AGENCY

The responsibility for environmental issues is that of the Environment Agency. It was formed by the Environment Act 1995 when the activities of the NWC were amalgamated with the responsibilities of other organisations to cover environmental protection of air and land as well as water. It is a non-departmental public body and is mainly sponsored by the Department for Environment, Food and Rural Affairs, and the National Assembly for Wales. It regulates by working as authorisers, advisers, inspectors and enforcers (Environment Agency, 2002).

The Agency "puts people and their environments at the heart of everything it does". It works in the following areas:

- Air Quality. It regulates the release of pollutants into the air for over 2,000 of the larger industrial processes.
- Conservation. It has a key role in conservation and ecology, especially along rivers and in wetlands.
- Fisheries.
- Flood. About 5 million people live in flood risk areas in England and Wales. The agency warns people about flooding and builds flood defences.
- Land Quality. It seeks to prevent land from becoming polluted and help to make it safe when it has become contaminated.
- Navigation. It is one of the major navigation authorities in charge of inland rivers, estuaries and harbours in England and Wales.
- Recreation. It works to ensure that inland and coastal waters, and the land around them is used for recreation.
- Waste. It regulates waste management through a system of licences and provides advice on waste management methods including waste minimisation.
- Water Quality. It is responsible for the quality of fresh, marine, surface and underground water and aims to avert the risk of water pollution.
- Water Resources. It aims to secure the proper use of water resources in England and Wales by national planning and the monitoring and issuing of water abstraction licences.

Internationally, the Agency contributes to the world-wide environmental goal of sustainable development, which has been defined as "development that meets the needs of the present without compromising the ability of future generations to meet their own needs" (Environment Agency, 2002). It is also a very key player nationally in England and Wales in working for an improved environment. As an example, in 2002 it persuaded the United Kingdom government to agree to the biggest ever [£6 billion] programme of water company investment in environmental improvement (Environment Agency, 2002).

The Environment Agency is divided into a number of offices and services around the country. It has a head office, eight national centres providing technical and scientific expertise to support key areas of its work such as the National Flood Warning Centre and the National Water Demand Management Centre. The agency also provides 22 national Services, such as the National Laboratory Service and the National Library and Information Service, and the work on the ground is done by 26 area offices. The Agency sees that it can contribute greatly to a better environment by disseminating environmental information and promoting environmental education in schools and professional institutions (Environment Agency, 2002).

15.5 United Kingdom Water Companies: Three Examples

In consequence of its history and geography there are differences in how the water industry has developed within the United Kingdom. Its international outlook that comes from its past has now manifested itself in the development of the water industry in England and Wales to be a major global player. Thames Water is now the 3rd biggest water supply provider in the world (Thames Water, 2002). Although England and Wales have chosen the model of having private firms own the infrastructure and operate water supply and sewage disposal services, in contrast to Northern Ireland, Scotland and other countries, the companies nevertheless have significantly different characteristics. These differences will be illustrated below by the description of three important water companies in England and the involvement of foreign firms. These differences will clearly show that it is no longer possible to talk simply about a water undertaking supplying a town or a region in England and Wales and the water undertaking looking for technical solutions to water supply problems. The development of water companies has gone well beyond this model and the companies are now very much commercial concerns involved in high finance, takeovers and mergers.

15.5.1 ANGLIAN WATER GROUP

The Anglian Water Group [AWG] is the water services and infrastructure management group based in the east of England. It is made up of six divisions: Anglian Water Services [AWS] and five infrastructure management businesses. AWS' basic area stretches from the River Thames to the River Humber Estuary and serves the needs of over five million industrial, commercial and domestic customers. AWS, the core business of AWG, is a regulated water, wastewater and environmental service provider for residential and commercial customers in the Eastern region and Hartlepool. AWS' water resources consist of: ground water sources, accounting for about 45% of the supply, abstracting mainly from the chalk and limestone aquifers in the north, south, and east of its supply area; surface sources impounded in pumped storage reservoirs, mainly Rutland Water and Grafham Water, which provide about 50% of the supply; and direct river abstractions accounting for the remaining 5% (Anglian Water Group, 2002).

When it recognised the growing demand from other sectors for its infrastructure-related skills and experience, AWG developed a range of services providing design, finance, creation and operational support to asset owning clients which are marketed through AWG Utility Services, [which provides infrastructure maintenance, renewals, new-build and specialist services to the gas, water, telecommunications and electricity sectors], AWG Government Services, AWG Project Management Services, AWG Developments and Commercial Services, and AWG International Services (Anglian Water Group, 2002).

15.5.2 THREE VALLEYS WATER

Three Valleys Water is the largest water-only supplier in the U.K., providing an average of 800 million litres of water per day to three million customers to the north, west and south of London. The sewerage and sewage disposal services within its area are provided by Thames Water. Three Valleys Water has been supplying water to the local community for more than 130 years through its predecessors, private companies Lee Valley Water, Colne Valley, and Rickmansworth, which combined forces in 1991 to form Three Valleys Water plc. The company grew again following a merger in October 2000 with its sister company North Surrey Water, which was formed in 1973 from four founder companies. The company is owned by Vivendi Water U.K., part of the international company Vivendi Environment which has about 215,000 employees in 100 different countries. Vivendi is based in France and was previously known as Compagnie General des Eaux [CGdE] (Three Valleys Water, 2002).

In England, the profits of the previous private water companies had been limited to a maximum 10% dividend on shares. CGdE was the first foreign company to recognise the fact that in the new privatised environment the control in the future would not be on profits but on prices. Therefore, if savings could be made by greater efficiency and economies of scale brought about by mergers then it reasoned that profits could be higher. Therefore, it bought Lee Valley and has acquired more water companies since.

In the early 2000s Vivendi decided to become a media giant, so when it purchased Universal Studios it became Vivendi Universal and was heavily debt laden. Therefore in November 2002, when it entered a battle with another telecom company for a third telecom company, it put its water businesses up for sale to raise cash. This put Three Valleys in the frame, as well as the Folkestone and Dover and Tendering Hundred water companies. It had already sold its interests in Bristol Water, a company which had been owned by a subsidiary of the infamous ENRON, and South Staffordshire Water.

In the same month, Vivendi's environmental arm made a bid to acquire Southern Water for £2 billion from First Aqua. A consortium of banks and investors was set up to acquire the company while Vivendi worked through the various regulatory hurdles. Southern Water supplies about 2.5 million people in the south east of England and provides sewerage systems and sewage treatment services to about 4 million people. However the move was blocked by the Government's Department of Trade and Industry [DTI] because the Competition Commission's view was that the acquisition "may be expected to operate against the public interest" by making it harder to compare prices across the market, a duty carried out by Ofwat. Therefore the DTI asked Ofwat and the Office for Fair Trading to examine a number of remedies to offset these concerns such as requiring Vivendi to sell its other water companies.

This saga demonstrates the principle, that Ofwat has always held to that there needs to be enough separate water companies in the Kingdom for it to compare the prices of all of them in order to ensure some form of competition within a monopolistic industry.

15.5.3 THAMES WATER

Thames Water supplies an average of about 3,000 million litres of water every day to some 7.5 million people within the River Thames catchment. It is also responsible for the sewerage and sewage treatment within its area. It has 2,700 employees and supplies an area which includes London and a large part of central southern England. Approximately 75% of the Company's water supplies are drawn from storage reservoirs that are mainly filled from surface water sources, principally the Rivers Thames and Lee. The remainder is drawn from ground water sources. Water from these sources is treated at 112 treatment works and distributed through 312 service reservoirs and about 32,000 km of mains. The company has always been at the forefront of innovation. An example is the 2,500mm diameter ring tunnel under London which connects its major water treatment works (Thanes Water, 2002).

Since the year 2000, Thames Water has been owned by the German company RWE, forming one of its four business groups. The others are electricity, gas and waste (Thames Water, 2002). Thames Water is the only one of the four which is not based in Germany and is the second biggest business unit after electricity of which Innogy is a part. Its customer base will have grown from 26 million to 70 million when the takeover of American Waterworks is completed. The acquisition of Thames, Innogy, and American Waterworks has transformed RWE from a German-centric business into an internationally-minded one. RWE now has electricity, gas, water and waste businesses in more that 120 countries with about 124,000 employees and revenues of about £40 billion in the financial year 2000/2001. The long-term business plan of Thames is geared to growth, primarily in the United States of America and in mainland Europe where it has acquired Iberdrola's Spanish water businesses, Pridesa and Ondaguna, through its subsidiary RWE Aqua (Thames Water, 2002).

15.6 The Future

Much has been learnt in the United Kingdom from the experience of how its water service has developed. The period since the Second World War has been one of progressive amalgamation and the development of dynamic new organisational structures which have let the Kingdom's water supply industry become a truly global player. In addition the Kingdom has almost thirty years of integrated catchment management experience. One cannot be positive about what will happen in the future within the Kingdom as governments of various political shades take and lose power and the desires of customers develop through time. However, the probability is high that the following features will be incorporated in the changes which are sure to happen:

- The U.K. companies will stay private and become even more dynamic. As a result they will grow, not only becoming even bigger players globally but also becoming more diverse, covering services over a wider range of activities than water supply and sewage disposal.
- The companies, encouraged by the Regulators, will put even more emphasis on the maintenance and development of the infrastructure.
- They will become much more commercial and be subject to the interest and therefore takeovers from large international companies. In their turn they will become more active in takeovers themselves.
- Globalisation will also be of benefit to the developing parts of the world because these companies, for a combination of commercial and altruistic reasons, will want to develop their activities in the developing nations. This should have the effect of improving the lot of people in these countries by not only having improved water supply and sanitation but also the economic growth which has been generated and made possible by improved water supplies.
- Regulation will as have to become more sophisticated to ensure that the companies are properly regulated in the interests of their customers while avoiding heavy and bureaucratic regulation that would stifle the commercial dynamism of the companies. Regulation needs to avoid discouraging long-term investment in the water industry. The Regulator will have to ensure that the contentious issue of cross subsidy[45] is better checked and regulated, especially with the involvement of so many foreign companies in the Kingdom's water supply industry.
- Within the Kingdom there will be a drive to ensure that there is even more non-bureaucratic integration within each river catchment, with more integration of land and water activities involving all stake holders, not just the water companies and their regulators, but local government, planning organisations and the customers themselves as they insist on an even higher quality of life. Techniques of valuing environmental and biodiversity benefits and disbenefits will have to be developed if optimal development decisions are going to be made.
- The larger more dynamic companies will require more effective and efficient solutions to water supply and sewage disposal problems; operating techniques; leakage prevention; innovative water and sewage treatment; and more efficient use of water resources. Therefore there will be more research and activity in the quest for more innovative solutions to problems. As water demands increase, due to increased per capita consumption rather that population increases, there should not be an increased use of the basic water resources.

Finally, as the Kingdom is part of the European Union, there will be increased pressures to conform to even more Europeanisation of water supply activities, namely the new European [water] Framework Directive. Hopefully this directive will be fairly applied throughout the Union so that some companies do not benefit to the detriment of others.

---

[45] A cross subsidy sets prices for some groups of customers above their cost of service while subsidizing other groups of customers with prices set below cost of service.

References

Anglian Water Group, 2002. www.anglianwater.co.uk
Drinking Water Inspectorate, 2002. www.dwi.gov.uk
Encyclopaedia Britannica,1996. CD Version 97.1
Environment Agency, 2002. www.environment-agency.gov.uk
Institution of Water Engineers, 1961. *Manual British Water Engineering*, 3rdEd.,W. Heffer, Cambridge.
King, H.B., 1992. *Water-The Book*, Quiller Press.
Northern Ireland Water Service, 2002. www.waterni.gov.uk
Office of Water, 2002.www.ofwat.gov.uk
Scottish Water, 2002. www.scottishwater.co.uk
Thames Water, 2002. www.thanes-water.com
Three Valleys, 2002. www.3valleys.co.uk
Twort, A.C., F.M. Law, F.W. Crowley, and D.D. Ratnayaka, 1994. *Water Supply*, Arnold Press, London.

Chapter 16

Concluding Remarks

Donald E. Agthe
Economic Consultant
R. Bruce Billings
University of Arizona
Nathan Buras
University of Arizona

Urban water is both a market commodity and a public good.

16.1 Economic and Engineering Challenges in Urban Water Supply

Groundwater overdraft and surface water depletion due to expanding populations and increased water use per capita is a major challenge to urban water providers. An equally severe challenge is prevention of further pollution of drinking water sources and the cleanup of already polluted sources. Part of the answer to both challenges is wastewater treatment to allow effective use of recycled water and implementation of effective prices and markets in controlling water demand. Security of supply, both with respect to multi-year fluctuations in precipitation and temperature and preventing terrorist attacks, is important and will require increasing resources. Provision of safe affordable drinking water and appropriate wastewater collection and treatment for burgeoning urban populations worldwide has so far remained chronically beyond the reach of many countries. Effective political support for technical, economic, and engineering solutions to these challenges is essential.

D.E. Agthe et al. (eds.), Managing Urban Water Supply, 257–269.

16.2 Groundwater Overdraft

A major problem faced by urban water providers is increasing demands for potable water due to increases in population and income. Chapters 12, 13, and 14 discuss this problem with relation to groundwater overdraft. Houston, Texas was found to have severe subsidence problems resulting from both oil extraction and groundwater overdraft. Arizona and Israel have groundwater problems that result from over pumping for both agriculture and urban development. Mexico City is well known for its severe subsidence and water supply problems.

Overpumping of an aquifer can lead to subsidence of the overlying land, although this relationship is not straightforward. Depending on soil type and other factors, water tables can be lowered many meters by excess pumping before any subsidence occurs. The boundary between the vadose zone[46] and the water table is flexible. Recharge in excess of pumpage can raise the water table and reduce the size of the vadose zone. This allows some aquifers to be used for water storage. Once subsidence occurs, however, it is generally irreversible.

In Arizona, the use of economic and technical solutions to solve the overdraft problem is complicated by grand-fathered water rights for agriculture, mining, Native American Tribes and private wells. Changes in laws are needed to restrict private wells, provide for orderly transfer of water between basins, better synchronize groundwater and surface water management, and encourage the use of recycled water. Israel, like many other countries, continues to experience conflicts over water use between urban areas, agricultural interests and industry. An even more serious problem in the long run which is beyond the scope of this book is the conflict among countries which assert competing claims for water from various rivers and lakes. Effective mutual agreements among countries are essential to minimize future conflicts.

Some of the problems surrounding groundwater overdraft and the coordination of the use of ground and surface water supplies stem from the development of hydrology as a science. Surface water hydrology and groundwater hydrology developed as separate specializations within the field of hydrology. It is now recognized that the two water sources are interrelated. Pumping groundwater affects stream flows much more than was originally thought. In Arizona, as in many other places, the laws regulating the use of groundwater are separate from those regulating surface water. This situation complicates the management of local and regional water supplies. Legislation should be brought up to date to recognize the best current understanding of the interactions between ground water and surface water.

16.3 Soft Path Water Management

Soft path water management refers to the use of improved water management techniques rather than large physical investments in new water infrastructure such as dams, canals, and well fields. It stresses more efficient water management to maintain

---

[46] The vadose zone is the area between the surface and the water table.

water quality and to ensure adequate water supplies. While not generally increasing the physical quantity of water available, soft path management techniques increase the economic quantity of water available. Efficiency in end use is stressed in applying these techniques, some of which are discussed in Chapter 4.

Examples of soft path management applied to residential customers may include raising price to reduce the quantity of water demanded, offering water bill rebates in desert cities to encourage desert landscaping in new and existing homes, updating building codes for new construction and resale properties to force installation of low flow toilets, showerheads and faucets, requiring swimming pool covers and limiting outdoor plantings. Much of chapter 3 is devoted to the use of price to reduce water demand in order to improve economic efficiency of water use. Chapter 7 suggests that knowledge of customer classes can aid in establishing appropriate prices and systems for recycled water.

Dual water systems, supplying both potable and reclaimed or other lower quality water can be an effective means of improving economic efficiency of water use in areas with sufficient industrial, commercial or irrigation demands. Non-potable water can also be used for residential toilet flushing, car washing, and landscape watering. Dual delivery systems are costly and would only be economically efficient in areas with relatively high water supply cost and either high housing densities or extensive landscape irrigation. Retrofitting older neighborhoods might be done when a regularly scheduled replacement of the water pipes is called for. Significantly higher prices for drinking water compared to recycled water, along with effective promotional activities, provides a strong incentive for the use of recycled water. Drinking water revenue could be used to subsidize the provision of recycled water. However, large and lower income families could be negatively impacted by a higher price for drinking water, suggesting the use of a lifeline rate or a subsidy program

Water conservation can be obtained by changing behavior, such as taking shorter showers, or by installation of water saving devices such as low-flush toilets. Knowing *how* to save water does not by itself result in water savings. The challenge for those utilities desiring reduced water demand is to institute effective incentives for conservation combined with the dissemination of information on how to save water. Numerous studies have shown that raising prices can be an effective incentive. Results of a study by Agthe and Billings (1996) suggest that increasing water prices encourage the installation of low flow toilets and low flow faucets in residences. Chapter 3 suggests that residential customers and apartment owners substitute these water saving devices and other conservation measures for water when price rises.

Price increases can also be combined with other means such as publicity to reduce water use. Some cities have achieved reduced water demand by offering free conservation kits to home and apartment owners. Studies of the effectiveness of these devices and programs include, among others, Witcomb (1990), Maddaus (1987), Morgan and Pelosi (1980) and Stein and Kilgore (1990). A rebate program in Mesa, Arizona which refunded part of the water bill when desert landscaping was installed was found to be a cost effective means of encouraging new home- builders and owners to install desert landscaping within the guidelines provided by the city (Agthe et al., 1986). However, the amount of money rebated was small enough that most high-income owners of expensively landscaped homes did not join this voluntary program. The insufficiency of this economic carrot may encourage local water authorities to introduce landscaping restrictions on new homes similar to the requirements for low flow toilets and faucets. Federal law in the United States prohibits sales of toilets that exceed a maximum water flow capacity of 1.5 gallons (5.7 liters). However, there were many anecdotal stories of people of means going to demolition sites of older buildings and to Canada to obtain older high-flow toilets when this regulation became effective. However, the quality of low flush toilets has improved over time, reducing this practice.

Business and industrial consumers represent a fruitful area for development of soft path management. These customers frequently use large volumes of water for landscaping, cooling, and industrial processes that do not require drinking quality water. Furthermore, they are frequently located in close proximity to each other because of industrial and commercial zoning, transportation facilities, and economies of agglomeration or conglomeration. Targeting industrial users as customers for reclaimed water reduces the cost of the infrastructure needed to provide the dual water system relative to a city-wide system. It is almost sure to be more cost efficient than providing a dual water system for the more widely scattered and lower volume per customer residential users. In Chapter 8, Renzetti points out that many industrial users have their own wells and/or treatment facilities. This situation should provide the water utility a means to trade its recycled water for access to the industrial enterprise's supply of drinking quality water to the extent that it is possible for the industrial user to substitute the lower quality water for drinking quality water. However, some legislative action or a demonstration by the water utility that the trade will not hurt the firm's product quality or processing equipment may be necessary to facilitate the trade.

Working in the area of environmental regulation, Smith (1999) found in many cases that a change in attitude of the firm's management toward the rules facilitated product and process changes that improved the environmental condition and that in most cases there was little or no net cost to the company. This is likely to be true for many cases where conservation of drinking quality water is the objective. Several of the recommendations made in Chapter 7 concerning the necessity for understanding individual customer's needs and customer classes' needs would aid in the negotiation process.

The goal of soft path management policies is to bring greater efficiency in water use by the end users and water providers. This does not imply that soft path management is necessarily either more or less expensive that hard path management in any specific situation[47]. Soft path has expenses associated with it. For a reclaimed water system, for example, allocations must be monitored, industrial and residential consumers must be convinced though advertising and demonstration that it is in their best interest to use lower quality water, rules must be enforced, dual water systems built, and negotiations to trade for drinking quality water must carried out. What is different is that the costs may be more diffused. The costs for dams, canals, reservoirs, and

[47] The soft path concept, at least partially, has arisen because of the increasing expenses associated with hard path solutions. When most of the best sites for dams, reservoirs, canals, and pipelines have been exhausted, the cost of these hard path options increases greatly. In this circumstance, soft path may be an economically attractive alternative.

pipelines are often borne at the state and national levels as was the planning for these hard path solutions to water shortage problems. With soft path, the cost of urban infrastructure, monitoring and administration is likely to be borne by the water provider or local government. When landscaping, production processes, and low flow residential devices are mandated as part of the soft-path solution, the cost is also transferred to the individual water consumer. While the consumer often paid for the hard path solutions in his or her tax bills, the consumer's water use habits are directly connected to many of the costs of soft path.

16.4 Availability of Water Service

Gleick (2000) argues that safe drinking water is a fundamental human right that more than one billion people do not have. He argues that it is the duty of governments and aid organizations to reduce the human suffering caused by inadequate drinking water supplies. Many of the world's major cities have been unable to extend adequate services to their growing populations, and in many cases the new urban dwellers have been unable to afford the cost of connecting to the supply network. Urban squatter areas often have neither piped potable water nor adequate sanitation services. Most people in these areas lack both the economic ability and political clout to induce municipal suppliers to meet their needs. The result is widespread use of unsafe water for household uses and the spread of water borne disease.

Privatization of water services in some areas, such as England, has resulted in the potential for new investment to bring water/sewer systems up to standard, as discussed in Chapter 15. On the other hand, privatization has often brought prices that preclude large numbers of poor people from receiving water service. The choice facing some urban areas may be poor quality water service at low prices which people can afford or high quality service that is beyond their means. In some countries, connection fees and associated costs prevent low-income people from connecting to the service in spite of the circumstance that they could afford to pay the commodity charge for water if they were connected to the system. A study of the situation of urban water in Bolivia by Israel (2001) found that many families were forced to rely on water delivered by mobile water sellers which is more expensive than available system water because they could not acquire the capital needed to join the formal water system. There are low-income suburbs in Juarez, Mexico that cannot afford the capital to provide water system infrastructure. Their solution to the water service problem is to rely on water delivered by truck from a water seller. To provide safe water for all of their citizens, nations need to transfer financial resources from more wealthy citizens to less wealthy citizens in the form of grants, long term low interest rate loans, and welfare programs. At the world level, high income countries working through organizations such as the World Bank could provide grants to build the needed infrastructure for proper water service in the less developed countries.

16.5 Security of Supply

Security of supply has several interpretations. Hydrologists and atmospheric scientists focus on maintaining a given water supply in the face of seasonal and annual fluctuations in precipitation and stream flows. Public health officials are concerned with protecting drinking water quality by proper treatment methods and protecting the quality of aquifers and surface supplies from sources of contamination. Public officials, administrators, and water providers are responsible for the security of the water system from possible terrorist attacks. Prevention of terrorist attacks, which include deliberate interruptions of supply by system destruction and/or contamination of water supply sources or water in urban pipelines, must be addressed by these civil authorities. Potential interruption of supply in wartime has been a challenge throughout recorded history, but modern urban concentrations and technology have exacerbated these threats in recent years.

The hydrologic problems of supply are well recognized and are often solved by known methods such as groundwater recharge, dams and reservoirs and improved methods of measuring groundwater supplies and forecasting weather and stream flows. These systems do not come without a cost. Forecasters and hydrologists must be trained and provided the proper equipment to produce accurate measurements. Dams, canals, and reservoirs that adjust seasonal availability of water are generally large scale publicly financed projects that are paid for by many taxpayers. In addition, environmental economic costs in terms of fish migration blockages, reduction of usable land covered by reservoirs and changes in water quality because of changes in flow are associated with dams and reservoirs. Groundwater recharge using imported water or recycled water is not a cheap solution. Recharge wells must be drilled and recycled water must be treated before recharge. Furthermore, rising groundwater tables may disturb land uses overlying the aquifer or the recharged water may be of lower quality than the existing groundwater.

Since a basic demand for urban water is for drinking, cooking, washing, and other domestic purposes, this water is generally treated to meet minimal drinking water standards to prevent disease and epidemics resulting from groundwater or upstream surface water contamination. Treatment plants are expensive and, in some cases the treatments (for example fluoride) are controversial. In Chapter 11, Levin points out that politicians, seeking to keep public expenditures and taxes low, often reject new more expensive but cost-effective treatments for drinking water. The desire to present a view of urban water as safe and inexpensive often discourages policy makers from advocating additional treatment even though the value of the potential reduction in water borne disease greatly exceeds its cost.

In a world of international turmoil and evolving technology, deliberate contamination of the water supply or destruction of the water supply system may become easier for potential terrorists. However, remote electronic surveillance of treatment facilities and pipelines can mitigate this threat. Improved security, including more frequent testing of water quality at the treatment plant and at random places in the system, will raise the cost of providing water. If water is brought in from a distance as in the case of the Central Arizona Project or the projects serving Southern California, the long miles of unguarded canals and pumping stations will be expensive to protect. Bio-terrorist security is a relatively new concern for water planners and providers.

16.6 Creating a Legal Environment for Urban Water Supply Management

In nations like Israel, where water sources are considered to be the property of the state, the legislature and, perhaps, local political leaders, make the decisions concerning water. We saw in Kislev's Chapter 14 that most of the decisions were made by the central government as to the division of water between urban and agricultural uses as well as among urban areas. The monopoly water company owned by the state sets most of the water policies. However, their policies are strongly influenced by the preferences of the managers of the monopolistic company and its unionized employees. Government failure and politics can interfere with proper urban water provision by the public sector just as much as monopoly ownership can interfere with efficient, equitable provision in the private sector.

In nations where there is a mix of government and private entities providing urban water, the legal structure should be carefully designed. The job of designing the legal structure falls to the legislature and the governmental agencies responsible for water supply and quality. Expensive litigation between the entities will result if errors are made in designing the laws governing the extraction and distribution of water.

A legislature yielding to pressure to exempt private wells from regulation, allowing agricultural interests to continue using groundwater when imported surface water is available and failing to enact aquifer protection laws can seriously harm the ability of local water management to provide adequate urban water service at its lowest cost. Furthermore, subsidence (in the case of groundwater) may result and the water provider may be forced to use expensive distant sources to make up for the water that is "wasted" by poorly thought out laws or "special interest" protection.

Lawmakers should realize that water is an economic good which is both scarce and useful. Therefore, its allocation generally produces both winners and losers. Water has a commodity value in itself that is often unrecognized by the law. In many parts of the world, including the Western United States, history is replete with water conflicts. Urban areas along the Nile, in the states that border the Colorado River in the Western United States and in many of the nations in the Eastern Mediterranean Region have conflicts over water use resulting from competition for the same water source. The early cost of obtaining water was merely the cost of digging a well or traveling to the local spring to draw out water. Later, the Romans built expensive aqueduct systems to deliver water to cities and agriculture. Today, water is a commodity that can sometimes be bought and sold. Israel, for example, is importing water from Turkey in huge tanker ships. While there must be some concern for water delivered for drinking and other domestic necessity uses, the vast majority of water is used as an input for the production of other goods and services, with by far the largest share in agriculture. Lawmakers should recognize the commodity value of water by facilitating markets and setting efficient and equitable rules for trading water rights in the market place.

Given the expanding world population and the relatively limited supply of potable water, legislation should be enacted to protect water sources. Special land zoning should be enacted to protect aquifers and their recharge zones from land uses that lead to pollution of the water source. Similarly, land uses near reservoirs can be limited so that the surface water in storage is protected from pollution. Some landowners may object that they should be free to use their land as they wish. However, real estate ownership is, in reality, only ownership of a bundle of rights granted by society to use the land. None of those rights should allow pollution of the groundwater or surface water so that others using the common pool resource will have their ability to use this water diminished. Still, developers, landowners, and others will argue that economic development justifies ignoring the pollution problem and the losses to other potential users of taking excessive amounts of water from the common pool. Legislatures and local governments must resist this pressure from what are frequently influential citizens and constituents who prefer short-term gain compared to the long-term loss associated with destruction of the water source.

Imported water and its attendant problems require special legal consideration. Frequently, laws exist that prevent water from being transferred from its geographic location to the urban area desiring to purchase the water. As pointed out in Chapter 13, the rural areas affected by water transfers to urban areas have good reasons for wanting to keep and control their water supplies. Water is important to economic development and local wildlife. Furthermore, the danger of over exploitation of the aquifer by any extractor to meet export demand will diminish the economic position of the other users of the common pool. Higher pumping costs and diminished ability to cope with drought conditions will be experienced by the other users.

One of the earliest major transfers of agricultural water to fuel urban growth occurred in 1913 when the city of Los Angeles, California, completed an aqueduct connecting the city with the Owens Valley where the city had purchased farms and water rights. Gottlieb and Fitzsimmons (1999) discuss the similarity of the economic depletion as a result of the transfers of water out of the Owens Valley in California, the areas near Tucson and Phoenix, and a special case involving the Metropolitan Water District in the Los Angeles area and the Castic-Lake Water Agency. They point out that each region had declining property tax roles, employment opportunities, and incomes after the transfers because of the decline in agriculture. They refer to the Owens valley as a "water colony" of the city of Los Angeles. It is our observation that the Castic Lake region (near Los Angeles) as well as the Tucson and Phoenix water transfer regions are sufficiently close to major cities that urban employment opportunities are available to the displaced agricultural employees. Furthermore, we contend that the urban growth occurring on the retired land might have happened anyway as the "highest and best use" of the land shifted from agricultural to urban. The Owens Valley, located approximately 150 miles (241 km) north of Los Angeles, does not have urban employment substitutes for the former agricultural employment available. Furthermore, lack of a sufficient amount of available water will discourage urbanization and other industries from locating in the valley. It is recommended that all transfers should be reviewed by knowledgeable individuals representing an agency of government that is higher than the level of government benefiting or losing from the water transfer. In New Mexico, inter-basin and intra-basin water trades require approval from a Water Master.

Corporate farms who regard water as one of their marketable assets probably are more willing to sell water rights to nearby expanding urban areas, with no concern for "third party" effects, particularly when crop prices are low. Gottlieb and Fitzsimmons (1999) point out that water transfers involving corporate farming cause especially severe problems since they often control the largest number of water rights in the basin and are driven by absentee owners to maximize profits. These transfers allow urban areas an opportunity to obtain water for their growing populations and economies while at the same time offering large corporate farms an opportunity to bail out by selling their valuable asset, water rights, to the urban water provider. While this may appear to be a win-win transaction, the future of the local economy is often jeopardized by the loss of water and the urban area has little incentive to employ water conservation practices if distant water can be imported at a low enough price.

In the Western United States water rights tend to belong with land. Thus, when Tucson Water purchased farms in the nearby Avra Valley to take advantage of their groundwater rights, it was discovered that the utility could not legally import the water to Tucson. Extensive litigation was necessary for Tucson to obtain the water associated with the purchased land.

In addition to the water rights issues associated with the distant resource pool, legislation in needed to define and protect the water during transfer and storage before its use. If a river is used to transport the water from the land where it was purchased, it must be resolved that landowners with water rights to the river water do not have the right to use the purchased water as a part of their water rights during a drought. Furthermore, these downstream landowners should also be prohibited from withdrawing this water for temporary use in a way that diminishes the quality of the water. There is little incentive for the urban area to purchase water from a distant source if its transportation cannot be reasonably assured.

Since the water supply available in any given season is usually different from water demand during that season, imported water frequently must be stored for future use in a following season or, perhaps, in following years. This storage could be in a local aquifer by using injection wells and/or a natural recharge basin. The storage may also be in a large reservoir that could also be used for recreational purposes. If the water provider storing this water shares a common storage aquifer or reservoir with other water providers, it must be given assurance that other water providers also pumping from that aquifer or reservoir will respect its rights to the recharged water (Agthe, 1986; Coe, 1979; Gleason, 1976). Furthermore, appropriate land zoning to protect the storage reservoir or aquifer is needed to insure water quality for the owner of the stored water. Otherwise, there is no incentive for any of the providers to develop and use such common storage systems.

Groundwater and, to a lesser extent, surface water is a common pool resource in which withdrawal in excess of recharge leads to a reduction of the water supply for all users. Withdrawal in excess of recharge diminishes the water supply and the aquifer storage capacity for all users. It also raises the costs of pumping water, water prices and therefore the economic cost of those products made using the water. The common pool problem is that when an aquifer is overlaid by more than one water user, all of the users have an incentive to pump as much water as they can use with little regard to the economic damages to the other pumpers and environmental damage to the aquifer. The incentive is for all users to pump as much as they can (as fast as they can) because what they don't pump will go to one of the other water extractors. The larger the group of potential water users and the more disparate their interests, the more severe this problem becomes. A small, closed group with homogenous circumstances can often successfully negotiate a mutually beneficial water management plan. Negotiation between disparate interests such as farmers and urban water utilities is less likely to lead to a mutually beneficial agreement. This is why rural residents fear the entry of a distant urban user into their aquifer or surface water resource. Regional basin management and laws governing pumping are needed to solve the *common pool problem.*

For many years, the relationship between surface water and groundwater was not well known. However, it is now clear that surface water flows affect water tables and groundwater extraction can influence river flows. Litigation between surface water users and well owners along the Colorado River is now taking place. The surface water users want many wells that are up to about one-half mile (800m) from the Colorado River shut down because they claim that the well owners are using water that rightfully belongs to them through their surface water rights. Many of these water rights are for water to serve urban areas in California and Arizona as spelled out in the Colorado River Compact. Much more is now known about the relationship between surface flows and underground flows through aquifers, and laws should be updated to reflect this knowledge.

Closely related to the need to develop a legal environment recognizing the relationship between groundwater and surface water is the need to develop legal procedures addressing the damages caused by subsidence related to excess pumping of groundwater. The legal implications of subsidence caused by overdrafting an aquifer are daunting. The University of Arizona Water Resources Research Center (2001) reported that retrieving damages for cracked foundations, sunken highways, broken utility lines, and leaky irrigation canals from those responsible for the damage might be very difficult as no legal basis for lawsuits has yet been established. Identifying those causing damage and the extent to which each entity is responsible is very difficult. These damages are also likely to cause numerous disputes between damaged parties and their property insurance companies. Thus, a significant cost of subsidence, in addition to physical damage, is the litigation needed to resolve damage claims against those entities responsible for the overdraft.

16.7 Wastewater Recycling

In the chapters concerning Arizona and Israel, it was pointed out that while recycling of treated waste water is increasing, much treated waste water is still merely "dumped into a stream" after leaving the treatment plant. In a recent study by Agthe (et. al. 2000) it was found that most of the city of San Diego, California's treated waste water was discharged into the Pacific Ocean and that much of the city of El Paso, Texas' treated waste water was discharged into the Rio Grande River. What makes these cases significant is that these areas, located in developed economic regions, are desert locations subject to acute water scarcity. Even though technology exists, these regions have not yet been able to market their treated recycled water or, in most cases, do much with recharging their local aquifers with this water. Dual water systems are expensive and subsidies are needed for initial customers to overcome the cost of creating the second water system needed to deliver reclaimed water. The irony is that, with standard accounting, recycled water is of lower quality but simultaneously more expensive than drinking quality water. This creates a marketing problem for the municipality. Better marketing of recycled water is needed to overcome the current user's preference for drinking quality water in those situations where recycled water can be used, leaving the potable water for those uses that require it.

16.8 Markets and Prices

In Chapter 6, Colby discusses the pros and cons of water markets, and concludes that in many cases the market for a particular source of water has too few buyers and/or sellers to be an effective market. In these cases compacts and agreements better serve bargainers. In a market of too few buyers and/or sellers the relative bargaining strength of the participants often determines the resulting prices and quantities. Monopoly or monopsony[48] prices and quantities can easily result. Gottlieb and Fitzsimmons (1999) point out that in any particular basin, water transactions frequently are negotiated exchanges among agencies and not effective markets with competitive bids among all users. Furthermore, third parties may have their welfare reduced by the actions of the bargaining parties.

Chapter 3 suggests that water rights could be held and traded by individual urban water customers. However, there is a cost to the water district of providing a trading mechanism, recording transactions, and enforcing property rights. Furthermore, the initial distribution of these water rights confers benefits on some individuals at the expense of other individuals. Determining this new welfare distribution is both legally difficult and politically hazardous. The concerns for the welfare of third parties and the possibility of monopoly or monopsony gains are similar to those discussed in Chapter 6.

Chapters 3 and 13 both discuss the need for keeping municipal water prices in line with inflation. Since the economy changes from year to year, this is necessary for revenue raising and for effective use of price to conserve water. Chapter 3 also discussed the relationship between income and urban water consumption. The suggestion of increasing block rates for residential water systems is based on the finding that increased income leads to increased water demand and that consumers with higher incomes have a better ability to pay for water. Thus, properly designed rate structures can be used as an effective part of a water conservation program. Increasing block rates can also facilitate financing the extension of piped water to low income urban areas.

[48] Monopoly is defined as a single seller of a commodity and a monopsony is a single buyer. Thus a water utility is often a monopoly as the only seller of retail water services in a designated area. A geographically isolated mine may be a monopsony in its labor market, since there are no other employers in the area.

Increasing block rates would let those with high incomes and water consumption help pay for water for those less fortunate.

16.9 Future Management—More Business Like?

Much of our book has been a call for more business like management of urban water supply. Almost all of the chapters cite the need for proper pricing of water. Chapter 7 called for understanding markets and the use of marketing techniques as water conservation tools. This chapter also advocates taking advantage of economies of scale at various steps in treatment, distribution, and administration of urban water supply.

It is easy to confuse business-like behavior with private business ownership and control of the urban water supply. Governments often take on responsibilities that are nonprofit or economic loss generating in nature. Urban water is a good candidate for government provision or, at least, government regulated private production. Water provision's public health value offers great external benefits to society at large that would not be counted in a private provider's profits. Thus, much of our reference to "business like" behavior relates to a government or regulated provider adoption of more business like practices to reduce the cost of water provision.

Barlow and Clarke (2002) argue that multinational companies raise the price of water so high that poor people can't afford it. Thus, privatization without effective regulation is not a good solution to the problem of achieving equity in the urban water market. Chapters 3 and 5 emphasized the need for equity in pricing and the difference in consumer behavior and affordability between rich and poor residents. Suggested remedies included increasing block rate structures and low income subsidies. However, the important point in establishing water rates is to achieve both affordability and equity along with conservation objectives.

16.10 The Role of Political Economy in Urban Water Management

The later chapters in this book addressing applied topics and descriptions of regional conditions have all implied that political considerations have interfered with effectively using some of the proposed solutions. Thus, these problems must be viewed from the standpoint of political economy as well as from standpoints of technical economics and engineering. Without political support, many appropriate technical and economic solutions to urban water supply problems cannot be implemented. Appropriate and efficient urban water provision requires political leaders with the will and ability to make decisions for the good of all. For many leaders, water supply improvement is a low priority in the absence of a crisis. Therefore, water system investments with long term benefits such as replacing leaking mains, extending service to low income residents, and enhanced water treatment are often ignored, even though their benefits would substantially exceed their costs.

Finally, we should make a clear distinction between the *value* of water, its *price* charged to the users and the *cost* of providing it in a developed urban environment.

Many publicly operated water utilities charge prices for water which do not cover all operating and capital costs. *Value* is often a subjective concept. For example, the value of a loaf of bread to a starving person is much higher than to a person who just had a full meal, although both could purchase the loaf of bread for the same price. Water has an intrinsic worth, which can be quantified only when it is related to the services it provides. As long as water of high quality is supplied in adequate quantity and at low prices in an urban area, the perception often is that it has little, if any, value. Then, the danger is that the regional water resources may be misused and abused.

The highest value of water is that it sustains human, plant and animal life now and in the future. Decision makers must consider future generations and the quality of the environment when providing and managing current urban water supplies.

References

Agthe, D.E., 1986. "Potential Benefits and Costs of In ground Storage of Imported Water," *Water Resources Bulletin*, 22(1): 129-131.

Agthe, D.E., M.W. Garcia, and L. Goodnough, 1986. "Economic Evaluation of a Rebate Program to Save Water," *J. of Environmental Systems,* 16(2): 81-86.

Agthe, D.E. and R.B. Billings, 1996. "Water-Price Effect on Residential and Apartment Low-Flow Fixtures," *J. of Water Resources Planning and Management*, 124(2): 113-117.

Agthe, D.E., R.B. Billings and J.C. Tinney, 2000. "Economic Considerations in Marketing Reclaimed Water," *Forum of the Association of Arid Lands Studies*, 16(1): 1-7.

Barlowe, M., and T. Clarke, 2002. *Blue Gold,* The New Press, New York.

Coe, J.L., 1979. "Groundwater Storage for California Water Project," *Journal of Irrigation and Drainage Division, ASCE,"* 105(3): 305-315.

Gleason, E., 1976. "Water Projects Go Underground," *Ecology Law Quarterly,* 5: 652-668.

Gleick, P.H., 1999. "The Human Right to Water." *Water Policy*, 1: 487-503.

Gottlieb, R. and M. Fitzsimmons, 1991. *Thirst for Growth: Water Agencies as Hidden Government in California,* University of Arizona Press, Tucson, Arizona.

Israel, D.K., 2001. "The Distributional Impact of Price Ceilings on Water in Developing Countries: An Analysis of Household Water Expenditures in Urban Bolivia," A paper presented at the Southern Economic Association, Tampa, Florida, November.

Maddaus, W.B., 1987. "The Effectiveness of Water Conservation Measures," *J. of the American Water Works Association*, 79(3): 52-58.

Maddaus, W.B., 1987. *Water Conservation*, American Water Works Association, Denver, Colorado.

Morgan, W.D. and P. Pelosi, 1980. "The Effects of Water Conservation Kits on Water Use," *J. of the American Water Works Association,* 72(3): 131-133.

Smith, R.S., 1998. *Profit Centers in Industrial Ecology,* Quorum Books, Westport, Connecticut.

Stein, S.M., and R. Kilgore, 1990. "Interaction between Water Conservation Measures and the Effects on Conservation Programs," *Proceedings of Conserve 90*, AWWA, Denver, Colorado.

University of Arizona Water Resources Research Center, 2001. "What Recourse is Available when Subsidence Damages Private Property," *Arizona Water Resource*, 9(3): 1.

Witcomb, J.B., 1980. "Water Use Reductions from Retrofitting Indoor Water Fixtures," *Water Resources Bulletin*, 26(6): 921-926.

Index

Water Science and Technology Library

1. A.S. Eikum and R.W. Seabloom (eds.): *Alternative Wastewater Treatment.* Low-Cost Small Systems, Research and Development. Proceedings of the Conference held in Oslo, Norway (7–10 September 1981). 1982 ISBN 90-277-1430-4
2. W. Brutsaert and G.H. Jirka (eds.): *Gas Transfer at Water Surfaces.* 1984 ISBN 90-277-1697-8
3. D.A. Kraijenhoff and J.R. Moll (eds.): *River Flow Modelling and Forecasting.* 1986 ISBN 90-277-2082-7
4. World Meteorological Organization (ed.): *Microprocessors in Operational Hydrology.* Proceedings of a Conference held in Geneva (4–5 September 1984). 1986 ISBN 90-277-2156-4
5. J. Němec: *Hydrological Forecasting.* Design and Operation of Hydrological Forecasting Systems. 1986 ISBN 90-277-2259-5
6. V.K. Gupta, I. Rodríguez-Iturbe and E.F. Wood (eds.): *Scale Problems in Hydrology.* Runoff Generation and Basin Response. 1986 ISBN 90-277-2258-7
7. D.C. Major and H.E. Schwarz: *Large-Scale Regional Water Resources Planning.* The North Atlantic Regional Study. 1990 ISBN 0-7923-0711-9
8. W.H. Hager: *Energy Dissipators and Hydraulic Jump.* 1992 ISBN 0-7923-1508-1
9. V.P. Singh and M. Fiorentino (eds.): *Entropy and Energy Dissipation in Water Resources.* 1992 ISBN 0-7923-1696-7
10. K.W. Hipel (ed.): *Stochastic and Statistical Methods in Hydrology and Environmental Engineering.* A Four Volume Work Resulting from the International Conference in Honour of Professor T. E. Unny (21–23 June 1993). 1994
    10/1: Extreme values: floods and droughts ISBN 0-7923-2756-X
    10/2: Stochastic and statistical modelling with groundwater and surface water applications ISBN 0-7923-2757-8
    10/3: Time series analysis in hydrology and environmental engineering ISBN 0-7923-2758-6
    10/4: Effective environmental management for sustainable development ISBN 0-7923-2759-4
    Set 10/1–10/4: ISBN 0-7923-2760-8
11. S.N. Rodionov: *Global and Regional Climate Interaction: The Caspian Sea Experience.* 1994 ISBN 0-7923-2784-5
12. A. Peters, G. Wittum, B. Herrling, U. Meissner, C.A. Brebbia, W.G. Gray and G.F. Pinder (eds.): *Computational Methods in Water Resources X.* 1994
    Set 12/1–12/2: ISBN 0-7923-2937-6
13. C.B. Vreugdenhil: *Numerical Methods for Shallow-Water Flow.* 1994 ISBN 0-7923-3164-8
14. E. Cabrera and A.F. Vela (eds.): *Improving Efficiency and Reliability in Water Distribution Systems.* 1995 ISBN 0-7923-3536-8
15. V.P. Singh (ed.): *Environmental Hydrology.* 1995 ISBN 0-7923-3549-X
16. V.P. Singh and B. Kumar (eds.): *Proceedings of the International Conference on Hydrology and Water Resources* (New Delhi, 1993). 1996
    16/1: Surface-water hydrology ISBN 0-7923-3650-X
    16/2: Subsurface-water hydrology ISBN 0-7923-3651-8

16/3: Water-quality hydrology ISBN 0-7923-3652-6
16/4: Water resources planning and management ISBN 0-7923-3653-4
Set 16/1–16/4 ISBN 0-7923-3654-2

17. V.P. Singh: *Dam Breach Modeling Technology.* 1996 ISBN 0-7923-3925-8
18. Z. Kaczmarek, K.M. Strzepek, L. Somlyódy and V. Priazhinskaya (eds.): *Water Resources Management in the Face of Climatic/Hydrologic Uncertainties.* 1996 ISBN 0-7923-3927-4
19. V.P. Singh and W.H. Hager (eds.): *Environmental Hydraulics.* 1996 ISBN 0-7923-3983-5
20. G.B. Engelen and F.H. Kloosterman: *Hydrological Systems Analysis.* Methods and Applications. 1996 ISBN 0-7923-3986-X
21. A.S. Issar and S.D. Resnick (eds.): *Runoff, Infiltration and Subsurface Flow of Water in Arid and Semi-Arid Regions.* 1996 ISBN 0-7923-4034-5
22. M.B. Abbott and J.C. Refsgaard (eds.): *Distributed Hydrological Modelling.* 1996 ISBN 0-7923-4042-6
23. J. Gottlieb and P. DuChateau (eds.): *Parameter Identification and Inverse Problems in Hydrology, Geology and Ecology.* 1996 ISBN 0-7923-4089-2
24. V.P. Singh (ed.): *Hydrology of Disasters.* 1996 ISBN 0-7923-4092-2
25. A. Gianguzza, E. Pelizzetti and S. Sammartano (eds.): *Marine Chemistry.* An Environmental Analytical Chemistry Approach. 1997 ISBN 0-7923-4622-X
26. V.P. Singh and M. Fiorentino (eds.): *Geographical Information Systems in Hydrology.* 1996 ISBN 0-7923-4226-7
27. N.B. Harmancioglu, V.P. Singh and M.N. Alpaslan (eds.): *Environmental Data Management.* 1998 ISBN 0-7923-4857-5
28. G. Gambolati (ed.): *CENAS. Coastline Evolution of the Upper Adriatic Sea Due to Sea Level Rise and Natural and Anthropogenic Land Subsidence.* 1998 ISBN 0-7923-5119-3
29. D. Stephenson: *Water Supply Management.* 1998 ISBN 0-7923-5136-3
30. V.P. Singh: *Entropy-Based Parameter Estimation in Hydrology.* 1998 ISBN 0-7923-5224-6
31. A.S. Issar and N. Brown (eds.): *Water, Environment and Society in Times of Climatic Change.* 1998 ISBN 0-7923-5282-3
32. E. Cabrera and J. García-Serra (eds.): *Drought Management Planning in Water Supply Systems.* 1999 ISBN 0-7923-5294-7
33. N.B. Harmancioglu, O. Fistikoglu, S.D. Ozkul, V.P. Singh and M.N. Alpaslan: *Water Quality Monitoring Network Design.* 1999 ISBN 0-7923-5506-7
34. I. Stober and K. Bucher (eds): *Hydrogeology of Crystalline Rocks.* 2000 ISBN 0-7923-6082-6
35. J.S. Whitmore: *Drought Management on Farmland.* 2000 ISBN 0-7923-5998-4
36. R.S. Govindaraju and A. Ramachandra Rao (eds.): *Artificial Neural Networks in Hydrology.* 2000 ISBN 0-7923-6226-8
37. P. Singh and V.P. Singh: *Snow and Glacier Hydrology.* 2001 ISBN 0-7923-6767-7
38. B.E. Vieux: *Distributed Hydrologic Modeling Using GIS.* 2001 ISBN 0-7923-7002-3

Water Science and Technology Library

39. I.V. Nagy, K. Asante-Duah and I. Zsuffa: *Hydrological Dimensioning and Operation of Reservoirs.* Practical Design Concepts and Principles. 2002 ISBN 1-4020-0438-9
40. I. Stober and K. Bucher (eds.): *Water-Rock Interaction.* 2002 ISBN 1-4020-0497-4
41. M. Shahin: *Hydrology and Water Resources of Africa.* 2002 ISBN 1-4020-0866-X
42. S.K. Mishra and V.P. Singh: *Soil Conservation Service Curve Number (SCS-CN) Methodology.* 2003 ISBN 1-4020-1132-6
43. C. Ray, G. Melin and R.B. Linsky (eds.): *Riverbank Filtration.* Improving Source-Water Quality. 2003 ISBN 1-4020-1133-4
44. G. Rossi, A. Cancelliere, L.S. Pereira, T. Oweis, M. Shatanawi and A. Zairi (eds.): *Tools for Drought Mitigation in Mediterranean Regions.* 2003 ISBN 1-4020-1140-7
45. A. Ramachandra Rao, K.H. Hamed and H.-L. Chen: *Nonstationarities in Hydrologic and Environmental Time Series.* 2003 ISBN 1-4020-1297-7
46. D.E. Agthe, R.B. Billings and N. Buras (eds.): *Managing Urban Water Supply.* 2003 ISBN 1-4020-1720-0

Kluwer Academic Publishers – Dordrecht / Boston / London